Studies in Computational Intelligence

Volume 739

Series editor

Janusz Kacprzyk, Polish Academy of Sciences, Warsaw, Poland
e-mail: kacprzyk@ibspan.waw.pl

The series "Studies in Computational Intelligence" (SCI) publishes new developments and advances in the various areas of computational intelligence—quickly and with a high quality. The intent is to cover the theory, applications, and design methods of computational intelligence, as embedded in the fields of engineering, computer science, physics and life sciences, as well as the methodologies behind them. The series contains monographs, lecture notes and edited volumes in computational intelligence spanning the areas of neural networks, connectionist systems, genetic algorithms, evolutionary computation, artificial intelligence, cellular automata, self-organizing systems, soft computing, fuzzy systems, and hybrid intelligent systems. Of particular value to both the contributors and the readership are the short publication timeframe and the worldwide distribution, which enable both wide and rapid dissemination of research output.

More information about this series at http://www.springer.com/series/7092

Sanaz Mostaghim · Andreas Nürnberger
Christian Borgelt

Editors

Frontiers in Computational Intelligence

 Springer

Editors
Sanaz Mostaghim
Faculty of Computer Science
Otto von Guericke University Magdeburg
Magdeburg
Germany

Christian Borgelt
Department of Computer and Information
  Science
University of Konstanz
Konstanz
Germany

Andreas Nürnberger
Faculty of Computer Science
Otto von Guericke University Magdeburg
Magdeburg
Germany

ISSN 1860-949X   ISSN 1860-9503   (electronic)
Studies in Computational Intelligence
ISBN 978-3-319-88487-5   ISBN 978-3-319-67789-7   (eBook)
https://doi.org/10.1007/978-3-319-67789-7

Printed on acid-free paper

This Springer imprint is published by Springer Nature
The registered company is Springer International Publishing AG
The registered company address is: Gewerbestrasse 11, 6330 Cham, Switzerland

Preface

Since the middle of the twentieth century, and accelerating sharply around the 1980s, computer science developed an area in which it is tried to equip computers with the capability to exhibit intelligent behavior, and thus to mimic the ability of humans and other intelligent animals to deal with uncertainty and vagueness, to learn from experience, and to adapt to changing environments. This area has aptly been named "Computational Intelligence" and nowadays belongs to the most actively researched areas not only in computer science, but even in the whole of science, technology and engineering.

Rudolf Kruse, to whom this book is dedicated on the occasion of his retirement from his post as a Professor at the Otto-von-Guericke-University of Magdeburg, Germany, helped substantially in shaping this area, not only with his own significant contributions, which are manifold, but also, and possibly even more, by promoting this area, by encouraging many others to enter it, and by supporting them to advance its state of the art.

This book collects several contributions that honor the achievements of Rudolf Kruse either directly or indirectly, by referring explicitly to his work or by showing the state of the art in specific areas of Computational Intelligence on which he had an influence. The main focus of these contributions lies on treating vague data as well as uncertain and imprecise information with automated procedures, which use techniques from statistics, control theory, clustering, neural networks, etc., to extract useful and employable knowledge.

The contribution by Enric Trillas and Rudolf Seising considers a problem that has been discussed in a sometimes heated fashion since fuzzy sets entered the stage of science, namely the interpretation and proper mathematical modeling of fuzzy sets. They remind us that despite a multitude of concepts with different names there are not many different types of fuzzy sets, but rather merely different formalizations of the linguistic phenomena of vagueness that one tries to model with fuzzy sets.

The contribution by Maria Ángeles Gil compares two definitions of fuzzy random variables, one of which originated from a book Rudolf Kruse wrote together with Klaus-Dieter Meyer ("Statistics with Vague Data"), which is contrasted to the one by Puri and Ralescu. Again, we find an interpretation problem at the heart of the distinction: Are fuzzy phenomena aspects of the real world or are they merely aspects of our observation of it? Depending on the answer, slightly different results and procedures are obtained.

The contribution by Inés Couso and Eyke Hüllermeier considers the case of statistically estimating ranking information from an incomplete observation of a ranking. This setting can be modeled along the general lines of handling set-valued information, similar to how set-valued information is treated in Dempster–Shafer theory or possibility theory, the latter of which was developed out of fuzzy theory as an alternative calculus to model uncertainty, orthogonal to probability theory and with a different emphasis.

The contribution by Thomas Runkler et al. takes a look at the area of fuzzy control, which has certainly been the most successful outcome of fuzzy theory w.r.t. actual applications. Again, the connection to Rudolf Kruse is very direct, as he played a decisive part in several applications of fuzzy control at Volkswagen (idle speed control, automatic gear shift, etc.). Thomas Runkler et al. study how a type-2 fuzzy set can be defuzzified, which is the necessary last step in the processing of a fuzzy controller based on type-2 fuzzy sets.

The contribution by Frank Klawonn enters the area of clustering, and especially fuzzy clustering in its various forms, on which Rudolf Kruse coauthored an influential book ("Fuzzy Cluster Analysis") with Frank Höppner, Frank Klawonn, and Thomas Runkler. Frank Klawonn considers dynamic data assigning assessment clustering, which was developed out of noise clustering, a fuzzy clustering approach to better deal with noise and outliers, to detect single clusters in an iterative fashion, and applies this approach to improve cluster detection in time-resolved data from the life sciences.

The contribution by Sara Mahallati et al. is also located in the area of clustering and considers the task of interpreting the structure of a clustering result, especially with the help of the well-known Dunn index, but also with visual assessment based on properly reordered distance matrices. The authors explore the close connection of both approaches to the time-tested single linkage hierarchical clustering method and apply their theory to the specific task of clustering waveform data, which, due to the time-dependent nature of the data, is closely connected to the preceding contribution.

The contribution by Malte Oeljeklaus et al. deals with one of the currently hottest topics in Computational Intelligence, namely deep learning neural networks for image analysis. With the discovery that the reach of the universal approximation theorem for neural networks is limited by the potentially needed huge size of a single hidden layer and the development of new activation functions and new training methods that allow, supported by advances in hardware, for efficient training of neural networks with many hidden layers ("deep" neural networks), the area of neural networks has seen unprecedented successes and consequently a surge of interest in recent years. This contribution deals with an application of such deep learning neural networks to traffic scene segmentation and recognition, which is a decisive step toward enabling autonomously driving vehicles ("self-driving cars").

The contribution by Christer Carlsson looks at the wider context and application potential of fuzzy methods and, more generally, soft computing technology in the area of management science and operations research or, as it is more often referred to today, business analytics. Here, fuzzy ontologies may be used for capturing domain-specific semantics for information retrieval by using fuzzy concepts, relations, and instances, and by defining and processing degrees of inclusion and coverage between concepts, which are then processed by a typical fuzzy min-max approach. However, since using software tools built with such sophisticated methodology requires high expertise of the operators, digital coaches are needed to help domain experts to fully exploit the benefits of such systems, which Carlsson also considers and advocates.

We are very grateful to all authors who accepted our invitation to contribute a chapter to this volume and to all reviewers who helped to improve the contributions. Furthermore, we express our gratitude to Janusz Kacprzyk, who made it possible to publish this book in the Springer series "Studies in Computational Intelligence." Last, but not least, we thank Springer-Verlag for the excellent collaboration that helped a lot to publish this book in time.

Magdeburg, Germany   Sanaz Mostaghim
Magdeburg, Germany   Andreas Nürnberger
Konstanz, Germany   Christian Borgelt
June 2017

Contents

What a Fuzzy Set Is and What It Is not?

Enric Trillas and Rudolf Seising

Abstract Although in the literature there appear 'type-one' fuzzy sets, 'type-two' fuzzy sets, 'intuitionistic' fuzzy sets, etc., this theoretically driven paper tries to argue that only one type of fuzzy sets actually exists. This is due to the difference between the concepts of a fuzzy set" and a "membership function".

1 Introduction

Although in the literature there appear 'type-one' fuzzy sets, 'type-two' fuzzy sets, 'intuitionistic' fuzzy sets, etc., this theoretically driven paper tries to argue that only one type of fuzzy sets actually exists. This is due to the difference between the concepts of a fuzzy set" and a "membership function". Both concepts deserve to be clarified.

Fuzzy sets can, for instance, be contextually specified by a membership function with values in the real unit interval but, nevertheless, membership functions with values out of this interval can be, in some situations, significant, suitable and useful. Situations in which either the range of their values cannot be presumed to be totally ordered, or it is impossible to precisely determine the membership numerical values, or the linearly ordered real unit interval produces a drastic simplification of the meaning of the fuzzy set's linguistic label by enlarging it through its 'working' meaning.

Indeed, this paper negates the existence of 'other fuzzy sets' than fuzzy sets, but it shows the possible suitability of designing their membership functions for

---

(*) To Professor Rudolf Kruse with the greatest esteem.

E. Trillas
University of Oviedo, Oviedo, Spain

R. Seising (✉)
Friedrich-Schiller University Jena, Jena, Germany
e-mail: R.seising@lrz.uni-muenchen.de

© Springer International Publishing AG 2018   1
S. Mostaghim et al. (eds.), *Frontiers in Computational Intelligence*, Studies
in Computational Intelligence 739, https://doi.org/10.1007/978-3-319-67789-7_1

sufficiently representing them, with as much as possible information available on the contextual behavior of the linguistic label. If, as Zadeh likes to say, 'fuzzy logic is a matter of degree', it is also a matter of design.

This paper starts with some historical notes on the development of the linguistic approach to fuzzy sets and fuzzy logic in the second half of the 1960s and in the first years of the 1970s. In the subsequent sections the paper presents a change of perspective by placing fuzzy sets in their natural domain, plain language; by going from 'general definitions' to 'design' in a given context, and depending on the meaning of the corresponding linguistic label inasmuch as a fuzzy set membership function should be carefully designed [19]. That is, they should be built up through a typical process of design, in which the simplicity of representation is a not to be forgotten practical value. Meaning is, indeed, only attributable to statements, and if simplicity is always considered as beautiful in science, 'design' is an art.

The main problem, to put it roughly, is representing words in a formal framework, similar to what Gottfried Wilhelm Leibniz had proposed more than 340 years ago. With his famous 'Calculemus!' he intended to resolve any differences of opinion: "The only way to rectify our reasonings is to make them as tangible as those of the Mathematicians, so that we can find our error at a glance, and when there are disputes among persons, we can simply say: Let us calculate [calculemus], without further ado, to see who is right." [6, p. 51].

We can find this idea of reducing reasoning to calculations already in the late 13th century in the work of the Catalonian, Ramon Llull. In his *Art Abreujada d'Atrobar Veritat* ("The Abbreviated Art of Finding Truth"), later published under the title *Ars generalis ultima* or *Ars magna* ("The Ultimate General Art" [Lull]. Leibniz had written his dissertation about Llull's *Art magna* and he named it "ars combinatoria" [7, p. 30].

Llull and Leibniz's arts have been steps on the plan for computing with concepts. All this deserves to be explained step-by-step; and in the first place, particularly the determination of words admitting of such a representation and where and by means of what is actually possible.

Remark What will not be taken into account in this paper are cases like that of the functions emerging from the aggregation of sets; that is, from aggregating their Characteristic Functions. For instance, if in the universe $X = \{1, 2, 3, 4\}$ the sets $A = \{1, 3\}$, and $B = \{1, 3, 4\}$ are aggregated by the mean $M(a, b) = (2a + 3b)/5$, what is obtained is not a set but the function $(2A + 3B)/2 = 1/1 + 0/2 + 1/3 + 0.6/4$, that is able to represent a fuzzy set provided a linguistic label for it (induced from those of A and B) can be known.

2   What Is It a Fuzzy Set?

2.1. Fuzzy Sets were launched in 1964 in three seminal papers by Lotfi A. Zadeh, a professor and chairman in the department of Electrical Engineering at the University of California, Berkeley. Zadeh construed his fundamental term of a "fuzzy set" without any non-mathematical meaning or application-oriented interpretation.

In the mid-summer of 1964 he was invited to a conference at Wright-Patterson Air Force Base in Dayton, Ohio. In his talk there, Zadeh considered problems of pattern classification, e.g. the process of representing the object patterns into a set of real variables which represent these patterns correctly and which would also be accepted by a computer.

Immediately after his travels he wrote on Fuzzy sets dealing with two problems:

- *Abstraction*—"the problem of identifying a decision function on the basis of a randomly sampled set", and
- *Generalization*—"referred to the use of the decision function identified during the abstraction process in order to classify the pattern correctly".

Zadeh first published this paper with his close friend Richard Bellman und Robert Kalaba as co-authors as a RAND-memo, and two years later in a scientific journal. Here he defined a "fuzzy set" as "a notion which extends the concept of membership in a set to situations in which there are many, possibly a continuum of, grades of membership." [2, 3, p. 1].

As a historically interested system theorist he had written the article "From Circuit Theory to System Theory" for the anniversary edition of the *Proceedings of the IRE* to mark the 50th year of the *Institute of Radio Engineers* in May 1962. In this article he stressed "the fundamental inadequacy of the conventional mathematics—the mathematics of precisely-defined points, functions, sets, probability measures, etc.—for coping with the analysis of biological systems, and that to deal effectively with such systems, which are generally orders of magnitude more complex than man-made systems, we need a radically different kind of mathematics, the mathematics of fuzzy or cloudy quantities which are not describable in terms of probability distributions" [22].

Two years later he had found this new mathematics and he explained its concepts in his second seminal paper: "Essentially, these concepts relate to situations in which the source of imprecision is not a random variable or a stochastic process but rather a class or classes which do not possess sharply defined boundaries" [23, p. 29].

In his third seminal paper, "Fuzzy Sets", he motivated the need for his new theory as follows: "More often than not, the classes of objects encountered in the real physical world do not have precisely defined criteria of membership. For example, the class of animals clearly includes dogs, horses, birds, etc. as its members, and clearly excludes such objects as rocks, fluids, plants, etc. However, such objects as starfish, bacteria, etc. have an ambiguous status with respect to the class of animals. The same kind of ambiguity arises in the case of a number such as 10 in relation to the "class" of all real numbers which are much greater than 1. Clearly, the "class of all real numbers which are much greater than 1," or "the class of beautiful women," or "the class of tall men, do not constitute classes or sets in the usual mathematical sense of these terms" [24].

2.2. After he had launched Fuzzy sets Zadeh proposed its use and applications to various disciplines. Computers and Computer Science (CS) have become part of Electrical Engineering (EE) Zadeh was very active to change his department's name from "EE" to "EECS". In 1969 in a talk at the conference "Man and Computer" in Bordeaux, France, he said: "As computers become more powerful and thus more influential in human affairs, the philosophical aspects of this question become increasingly overshadowed by the practical need to develop an operational understanding of the limitations of the machine judgment and decision making ability" [26, 27, p. 130]. He called it a paradox that the human brain is always solving problems by manipulating "fuzzy concepts" and "multidimensional fuzzy sensory inputs" whereas "the computing power of the most powerful, the most sophisticated digital computer in existence" is not able to do this. Therefore, he stated that "in many instances, the solution to a problem need not be exact", so that a considerable measure of fuzziness in its formulation and results may be tolerable. The human brain is designed to take advantage of this tolerance for imprecision whereas a digital computer, with its need for precise data and instructions, is not" [26, 27, p. 132]. He intended to push his theory of fuzzy sets to model the imprecise concepts and directives: "Although present-day computers are not designed to accept fuzzy data or execute fuzzy instructions, they can be programmed to do so indirectly by treating a fuzzy set as a data-type which can be encoded as an array [...]" [26, 27, p. 132].

2.3. Already in 1968 Zadeh has presented "fuzzy algorithms" [25]. Usual algorithms depend upon precision. Each constant and variable is precisely defined; every function and procedure has a definition set and a value set. Each command builds upon them. Successfully running a series of commands requires that each result (output) of the execution of a command lies in the definition range of the following command, that it is, in other words, an element of the input set for the series. Not even the smallest inaccuracies may occur when defining these coordinated definition and value ranges. Zadeh now saw "that in real life situations people think [...] like algorithms but not precisely defined algorithms". Inspired by this idea, he wrote: "The concept in question will be called fuzzy algorithm because it may be viewed as a generalization, through the process of fuzzification, of the conventional (nonfuzzy) conception of an algorithm. [25] To illustrate, fuzzy algorithms may contain fuzzy instructions such as:

(a) "Set y *approximately equal to* 10 if x is approximately equal to 5," or
(b) "If x is *large*, increase y by *several* units," or
(c) "If x is *large*, increase y by *several* units; if x is *small*, decrease y by *several* units; otherwise keep y unchanged."

The sources of fuzziness in these instructions are fuzzy sets which are identified by their names in italics.

To execute fuzzy algorithms by computers they have to receive an expression in fuzzy programming languages. Consequently, the next step for Zadeh was to define fuzzy languages. "All languages", he wrote in a paper on Architecture and Design of Digital Computers, "whether natural or artificial, tend to evolve and rise in level through the addition of new words to their vocabulary. These new words are, in effect, names for ordered subsets of names in the vocabulary to which they are added." [28, p. 16] He argued explicitly for programming languages that are—because of missing rigidness and preciseness and because of their fuzziness—more like natural languages. He mentioned the concept of stochastic languages that was published by the Finnish mathematician Paavo Turakainen in the foregoing year [21], being such an approximation to our human languages using randomizations in the productions, but however, he preferred fuzzy productions to achieve a formal fuzzy language. With his Ph. D student Edward T.-Z. Lee he co-authored a short sketch of a program to extend non-fuzzy formal languages to fuzzy languages in "Note on Fuzzy Languages" [5].

2.4. On the other hand, in the first years of Fuzzy sets Zadeh believed in successful applications of his new concepts in non-technical fields as he wrote in 1969: "What we still lack, and lack rather acutely, are methods for dealing with systems which are too complex or too ill-defined to admit of precise analysis. Such systems pervade life sciences, social sciences, philosophy, economics, psychology and many other "soft" fields" [25].

His search for application fields led to a period of interdisciplinary scientific exchange on the campus of his university between himself and the mathematician Hans-Joachim Bremermann, the psychologist Eleanor Rosch (Heider) and the linguist George Lakoff.

It was in these 1970s when psychologist Rosch developed her prototype theory on the basis of empirical studies. This theory assumes that people perceive objects in the real world by comparing them to prototypes and then subsequently ordering them. In this way, according to Rosch, the meanings of words are formed from prototypical details and scenes and then incorporated into lexical contexts depending on the context or situation. It could therefore be assumed that different societies process perceptions differently depending on how they go about solving problems [12]. When Lakoff heard about Rosch's experiments, he was working at the Center for Advanced Study in Behavioral Sciences at Stanford. During a discussion about prototype theory, someone there mentioned Zadeh's name and his idea of linking English words to membership functions and establishing fuzzy categories in this way. Lakoff and Zadeh met in 1971/72 at Stanford to discuss this idea after which Lakoff wrote his paper "Hedges: A Study in Meaning Criteria and the Logic of Fuzzy Concepts" [4]. In this work, Lakoff employed "hedges" (meaning barriers) to categorize linguistic expressions, he used the term "fuzzy logic" in his article and he therefore deserves credit for first introducing this expression in scientific literature. Based on his later research, however, Lakoff came to find that fuzzy logic was not an appropriate logic for linguistics: "It doesn't work for real natural languages, in traditional computer systems it works that way." [4]

However, "Inspired and influenced by many discussions with Professor G. Lakoff concerning the meaning of hedges and their interpretation in terms of fuzzy sets," Zadeh had also written an article in 1972 in which he contemplated "linguistic operators", which he called "hedges": "A Fuzzy Set-Theoretic Interpretation of

Hedges". Here he wrote: "A basic idea suggested in this paper in that a linguistic hedge such as very, more, more or less, much, essentially, slightly etc. may be viewed as an operator which acts on the fuzzy set representing the meaning of its operand [31].

2.5. Zadeh's occupation with natural and artificial languages gave rise to his studies in semantics. This intensive work let him to the question "Can the fuzziness of meaning be treated quantitatively, at least in principle?" [29, p. 160]. His 1971 article "Quantitative Fuzzy Semantics" [30] starts with a hint to these studies: "Few concepts are as basic to human thinking and yet as elusive of precise definition as the concept of »meaning«. Innumerable papers and books in the fields of philosophy, psychology, and linguistics have dealt at length with the question of what is the meaning of »meaning« without coming up with any definitive answers." [29, p. 159]

Zadeh started a new field of research "to point to the possibility of treating the fuzziness of meaning in a quantitative way and suggest a basis for what might be called quantitative fuzzy semantics" combining his results on fuzzy languages and fuzzy relations. In the section "Meaning" of this paper, he set up the basics: "Consider two spaces: (a) a universe of discourse, U, and (b) a set of terms, T, which play the roles of names of subsets of U. Let the generic elements of T and U be denoted by x and y, respectively. Then he started to define the meaning $M(x)$ of a term x as a fuzzy subset of U characterized by a membership function $\mu(y|x)$ which is conditioned on x. One of his examples was: "Let U be the universe of objects which we can see. Let T be the set of terms white, grey, green, blue, yellow, red, black. Then each of these terms, e.g., red, may be regarded as a name for a fuzzy subset of elements of U which are red in color. Thus, the meaning of red, M (red), is a specified fuzzy subset of U."

In his "Outline of a new approach to the analysis of complex systems and decision processes" [32] and in the three-part article "The concept of a Lingustic Variable and its Application to Approximate Reasoning" [33], in "Fuzzy Logic and Approximate Reasoning" [34] and finally in "PRUF—a meaning representation language for natural languages" [35] Zadeh developed a linguistic approach to Fuzzy sets. He defined linguistic variables as those variables whose values are words or terms from natural or artificial languages. For instance, "not very large", "very large" or "fat", "not fat" or "fast", "very slow" are terms of the linguistic variables size, fatness and speed. Zadeh represented linguistic variables as fuzzy sets whose membership functions map the linguistic terms onto a numerical scale of values.

2.6. A fuzzy set is a concept associated by a linguistic label of which no mathematical axiomatic theory is currently known. A fuzzy set is nothing other than something just exhibited in plain languages, and through the usual forms of speaking; it is a concept actually well anchored in language, and is useful for roughly, economically, and quickly describing what once perceived by people is translated into words. At each context, a fuzzy set is seen by the speakers as a unique entity associated to its linguistic label's use that, usually, is context-dependent and purpose-driven.

The usual confusion between fuzzy set and membership function in a given universe of discourse is not sustainable, since there is not a one-to-one correspondence, but a one-to-many between fuzzy sets and its membership functions that can be given by expressions of different functions, even sharing some common properties among them. As it is well known, a linguistic label P in a universe of discourse X does not admit of a single membership function, but several such functions can be chosen, unless P is a precisely used label.

For instance, the use of the label 'big' in the closed interval [0, 10] and in plain language, can be represented by several membership functions like, for instance, $x/10$, $x^2/100$, etc., depending on the available additional information on its shape. It is commonly accepted that the use of 'big' in the universe [0, 10] can be described by the four rules:

(1) x is less big than y $\iff$ x $\leq$ y;
(2) 10 is totally big;
(3) 0 is not at all big;
(4) If x can be qualified as big, it exists $\varepsilon(x) > 0$, such that all the points in the interval $[x - \varepsilon(x), x]$ can also be qualified as big.

With them, the membership functions representing 'big' can be all those mappings m_{big}: [0, 10] $\to$ [0, 1], such that:

(1') If $x \leq y$, then $m_{big}(x) \leq m_{big}(y)$;
(2') $m_{big}(10) = 1$;
(3') $m_{big}(0) = 0$;
(4') m_{big} is continuous.

There are an enormous amount of all the strictly non-decreasing functions [0, 10] $\to$ [0, 1] joining the points of coordinates (0, 0) and (1, 10). Anyway, the four laws (1')–(4') cannot specify a single m_{big}; for specifying one of them, some additional information is required. For instance, provided it can be presumed that function m_{big} should be linear, then it only exists $m_{big}(x) = x/10$, but if what can be supposed is that m_{big} is quadratic several of them exist, with $m_{big}(x) = x^2/100$ among them. Notice that provided it were known that the curve defined by m_{big} passes through the point (5, 0.6), then neither $x/10$, nor $x^2/100$, are acceptable. Etc.

Clearly, fuzzy sets do not admit to being specified by a single membership function, with the only exception corresponding to the case in which the linguistic label is precise, that is, its use in X is describable by 'if and only if' rules. With this exception, fuzzy sets cannot be confused with the measures of their meaning, or membership functions. For instance, provided the use of 'big' in [0, 10] were described by just the precise rule,

$$'x \text{ is big} \iff 8 \leq x \leq 10',$$

then its only membership function will be,

$$m_{big}*(x) = 0, \text{ if } 0 \leq x < 8, \text{ and } m_{big}*(x) = 1, \text{ if } 8 \leq x \leq 10;$$

and, accordingly with the 'specification axiom' of set theory [4], 'big' is specified by the set [8, 10] $= (m_{big}*)^{-1}(1)$.

It should be pointed out that function $m_{big}{}^*$ verifies the former properties (1′), (2′), and (3′), but not property (4′) as a consequence of the failing of rule (4), since for instance, for any $\varepsilon > 0$, no point 8—ε can be qualified as big in the current use of this label. Function $m_{big}{}^*$ is not a continuous one. It is rule (4) which distinguishes the imprecise from the precise uses of 'big' in [0, 10].

Notice that even defining the fuzzy set 'big' by each pair (big, m_{big}) instead of by only a function m_{big}, as it is also usually done [11], it not only follows that the linguistic label 'big' generates several fuzzy sets instead of a single one, but also a partial externalization of the concept of fuzzy set from language. The meaning of the imprecise word 'big', and what the functions m_{big} mean is hidden. For scientifically domesticating fuzzy sets, meaning and its measuring should be analyzed.

2.7. A linguistic label P names a property p the elements of the universe of discourse X enjoy, and which use is exhibited by the meaning attributed to the elemental statements 'x is P', for all x in X. Notice that the elements x can be physical or virtual, etc., but the new elements 'x is P' are just statements in the intellect, and belong to the set X [P] = {x is P; x in X} that is different of X; actually, meaning is attributed to statements, and it is usually context-dependent and purpose-driven. Meaning depends on the context on which the statements 'x is P' are used either in written form, or uttered, or gestured, etc., and depend on the purpose for such use [16, 17]. For instance, the same word P = odd, when used in the context of Arithmetic has a different meaning than when it is used in a social context, and, if in the first case it can only be used with purposes limited by the definition 'the rest of its division by two is one', in the second, it can be used with several and open purposes such as the descriptive, the insulting, etc. In the first case, its use is precise or rigid, but in the second it is imprecise or flexible. Analogously, in different contexts the word 'interesting' can be used with non coincidental purposes. Semantics is what really matters in language; without capturing the meaning of words, language is unintelligible.

The meaning of a word P is privative of a given universe of discourse; for instance, the same person can be tall in a population of pygmies, and short in one of giants. The concept of the meaning requires the joint consideration of both a universe of discourse X, and the word P, as it was formerly shown with the same word 'odd' in the respective universes of integer numbers, and of people.

2.8. To capture what it means P in X, it is necessary to know the relationship 'x is less P than y', expressing the recognition that x shows the property p named P less than y shows it [17, 18]; how the application of P varies along X. Of course, if P is precisely used in X, such a relationship just degenerates into 'x is equally P than y', and 'x is not equally P than y'. For instance, the numbers 3 and 21, 515 are equally odd, and the numbers 3 and 20 are not equally odd; in the universe of integer numbers there are only odd and not odd numbers.

Let's symbolically represent such relationship by 'x $<_P$ y'; it can be equivalently said that 'y is more P than x'. The symbol $<_P$ reflects a mathematical relation in X, of which it only can be quietly asserted that is reflexive, x $<_P$ x, for all x in X; a property assuring that the relation $<_P \subseteq X \times X$ is not empty. The graph (X, $<_P$) specifies the 'qualitative meaning' of P in X, and when the use of P is precise, the relations $<_P$ and $<_P^{-1}$ (whose intersection gives the relation $=_P$, 'equally P than'), just collapses in $=_P$. The qualitative meaning of a precisely used word P is the graph $(X, =_P)$, between whose arcs are the loops at each element in X, and, of course, if P is imprecise in X, the graph $(X, <_P)$ contains the graph $(X, =_P)$.

These graphs agree with the common view that when telling something people introduce some organization, or rough order, in what is taken into account.

By another side, it should be observed that, in plain language, words do 'collectivize' in the universe of discourse. For instance, if X is the set of London's inhabitants, P = young facilitates in X the linguistic-collective of the 'Young Londoners'. Linguistic-collectives, such as 'ripe tomatoes', 'high mountains', 'comfortable chairs', 'tall people', etc., are well anchored in plain languages since its speakers not only easily use them, but immediately capture what they express. Such 'collectives' are linguistically generated in X by the corresponding meaning of P in X, and, by using an old philosophical expression, it allows us to say that the linguistic label has some 'extension' in the corresponding universe of discourse.

Of course, linguistic-collectives are not always sets; they are only sets if P is precise, as it is, for instance, in the collective of the 'thirty five year old Londoners', as it is guaranteed by the 'axiom of specification' of the theory of sets [4]. Even if they are a kind of cloudy entities, linguistic-collectives are empirically recognized, they exist in language; they are 'a reality' in language like clouds are in the atmosphere.

A fuzzy set in X with linguistic label P, is nothing else than the linguistic-collective generated by P in X; it can be specified by the graph $(X, <_P)$ if P is imprecise, or the simpler graph $(X, =_P)$ if P is precise. There is no difference between the concepts of linguistic-collective, fuzzy set, and qualitative-meaning, they just denote the same concept.

2.9. Once a qualitative meaning of P in X is recognized, it should be pointed out that its defining relation $<_P$ is not always a linear one; that is, there are often elements x and y in X such that it is neither $x <_P y$, nor $y <_P x$; elements that are 'not comparable' under $<_P$. Typical examples are obtained with P = interesting in a universe of possible businesses, with P = beautiful in a universe of paintings, with P = odd in a universe of people, with P = nice with houses, etc. In collections of paintings for instance, there are often pairs of which it is impossible to state that any one of them is less, or more beautiful than the other.

Once the qualitative meaning $(X, <_P)$ is captured, it can be said that P is 'measurable' in X, since measures m_P of the extent up to which each x is P can be defined analogously to the former case of P = big in X = [0, 10]. If no relation $<_P$ can be even imagined, P is 'meaningless', or 'metaphysical' in X [17]. For instance, if $<_{big}$ is recognized as the linear order $\leq$ of the interval [0, 10] when such a word is used in it, there is no way of knowing $<_{big}$ seems to exist when 'big' is applied to dreams; hence 'big' is measurable in [0, 10], but it is in principle meaningless among dreams.

Although measurability is truly important for a scientific domestication of concepts, the metaphysical ones cannot be fully contemptible since at least they can have a true 'suggestive power'. What can be measured belongs to reasoning, and what cannot belongs to thinking; reasoning is but an organization of thinking for directing it towards a goal. A word that is currently metaphysical in X, can suggest to be applied in Y (different of X) in a measurable way. Anyway, this topic belongs to the kind of questions that are beyond what this paper is trying to consider.

3   How Fuzzy Sets Can Be Computationally Managed?

3.1. In the first place, it should be pointed out that a general definition of what is a measure should be liberated from the typical additive law always presumed in probability theory. It supposes that 'things grow' by superposition of non-overlapping pieces, and that the total measure is the sum of the measures of such pieces; something that cannot be always presumed, and less again with cloudy entities resembling linguistic-collectives.

From the background of game theory and decision-making, the Japanese physicist and engineer Michio Sugeno also had the idea that the property of additivity seemed to be too strong and therefore he reduced the integral form to monotonicity. In a later interview he recalled: "I put the adjective 'fuzzy' to this monotone measure simply because max-min-operations were used in its integral form as in fuzzy sets; this naming was later found to be not adequate. More precisely, I should have called it monotone measure or even 'non-additive measure'. I found that the monotonicity of the fuzzy measure well fits the calculations of max-min." Sugeno gave a mathematical foundation to this "fuzzy integral", it was later called the "Sugeno integral", at first in a Japanese journal in 1972 and in 1974 published in his doctoral thesis "Theory of Fuzzy Integrals and its Applications" [13–15].

Adapting Sugeno's concept of a 'fuzzy measure' [15] allows a general clear enough definition of a measure of meaning as follows [17, 18, 20].

Given the qualitative meaning $(X, <_P)$, a measure of it is a mapping $m_P: X$ [0, 1], such that:

(1) $x <_P y => m_P (x) \leq m_P (y)$,
(2) If z is maximal relatively to $<_P$, it is $m_P (z) = 1$,
(3) If z is minimal relatively to $<_P$, it is $m_P (z) = 0$.

This definition deserves some comments. Concerning (1), it just reflects that the growing variation of P along X, expressed by $<_P$, is translated into the growing of numbers in the unit interval given by its linear order $\leq$; what it does not reflect is how this growing is produced, something that each type of problem would require a particular form of decomposing elements in constitutive pieces.

Concerning (2), z is a maximal provided no other $x \in X$ exists such that $z <_P x$; a maximal is a 'prototype of P' in X, and its existence cannot be taken for granted, but if a single one exists it is called the maximum. Concerning (3), z is a minimal provided no other y exists such that $y <_P z$; a minimal is an 'anti-prototype of P' in

X, and its existence is also unsure, but if a single one exists it is called the minimum. When no maximal, or no minimal exist, then laws (2) or (3) cannot be applied; anyway, it does not imply that elements in X with measure one or zero are inexistent, such elements, if existing, can be respectively called 'working proto-types', and 'working anti-prototypes'. For instance, it is well known in probability theory there can be elements in the sigma-algebra of events that, not being an empty event, have a zero probability.

Nevertheless, the existence of prototypes and anti-prototypes seem to make consistent the recognition of the qualitative meaning; they allow the comparison of other elements in a form similar to the old Standard Meter Prototype for the (decimal system's) measuring of length. For instance concerning P = tall in a population, once recognized that Ruth is the tallest among inhabitants (Ruth is a prototype of tall), the tallness of the others is recognized by comparing them with Ruth; moreover, a stick with the same height as Ruth can serve for attaching relative and fractional measures of tallness. Of course, the same can be said once it is recognized that John is a less taller inhabitant, that he is an anti-prototype of tall in the population under consideration. For instance, in X = [0, 10], and respect to the former toy-example with P = big, there is initially recognized a unique maximal (the maximum 10), and a unique minimal (the minimum 0).

It should be pointed out that in the original Sugeno's definition of a 'fuzzy measure', X is a power-set 2^Ω, and the comparable relation (<) is the inclusion of sets $\subseteq$ with the maximum Ω, and the minimum $\emptyset$.

3.2. Let us recall that the three laws of a measure do not allow us to specify a single one unless P is precisely used in X; for instance, in the case of 'big' it was necessary to add some contextual information, or some reasonable hypothesis on the shape of the curve $y = m_{big}(x)$, as it could be, respectively, that it passes through (5, 0.6), or that it is linear, or quadratic, etc. If P is imprecise there is not a single measure for the meaning of P in X, but rather a set of them. It is similar to what happens, for example, with the probabilities of getting 'n points' ($1 \leq n$ 6) by throwing a single die, or with Sugeno's fuzzy measures containing many types of them as they are the big family of additive, sub-additive, and super-additive lambda-measures [15].

Only in the precise case is the measure unique. Since it is,

$$\text{If } x =_P y \Leftrightarrow x <_P y \ \& \ y <_P x => m_P(x) \leq m_P(y) \ \& \ m_P(y) \leq m_P(x) \Leftrightarrow$$
$$m_P(x) = m_P(y),$$

the measure preserves the relation $=_P$ that, in addition to reflexive is also symmetrical; provided $<_P$ were transitive, then $=_P$ is also transitive and, thus, an equivalence. Hence, because X is perfectly classified in the set specifying P (containing all the prototypes), and its complement (containing all the anti-prototypes), it only exists in the measure given by $m_P(x) = 1$ for all the first, and $m_P(x) = 0$ for all the second.

Notice that the only set with neither prototypes, nor anti-prototypes, is the empty set; it is a very 'odd set' since it is self-contradictory, $\emptyset \subseteq X = \emptyset^c$; indeed, it is the only self-contradictory set: $A \subseteq A^c \iff A = \emptyset$. Its acceptance as a set derives at least, from two practical reasons; the first is for guaranteeing that the intersection of sets is always a set, and the second for denoting the 'extensionality' of statements that are non-applicable to a given universe, and as it is, for instance, 'getting nine' in the universe of the six elemental events $\{1, 2, 3, 4, 5, 6\}$ obtainable when throwing a single die.

Hence, it can be said that the linguistic-collective given by the graph $(X, <_P)$, shows several contextual states $(X, <_P, m_P)$; each time a measure is specified, a 'contextual informational state' (in short, 'state') of the fuzzy set is manifested. Measures, or membership functions, represent known states of the fuzzy set; once designed, a membership function is practically employed to 'describe' the corresponding fuzzy set; they facilitate the management of fuzzy sets for computational purposes.

3.3. Once a measure m_P is specified, it defines the new relation in X,

$$x \leq_{mP} y \Leftrightarrow m_P(x) \leq m_P(y),$$

that is a linear ordering in X, and is obviously greater than $<_P$:

$$x < y => m_P(x) \leq m_P(y) \Leftrightarrow x \leq_{mP} y; \text{ that is, } <_P \subseteq \leq_{mP}.$$

If there can be elements x and y for which it is neither $x <_P y$, nor $y <_P x$, nevertheless one of the two numbers $m_P(x)$, $m_P(y)$ will be greater than the other; there are no incomparable elements under $\leq_{mP}$. The difference set $\leq_{mP} - <_P$ is not always empty, even not necessarily if $<_P$ is linear.

Thus, the new and linear relation $\leq_{mP}$ enlarges the qualitative meaning, and gives the linear 'working' meaning $(X, \leq_{mP})$, only known after a measure m_P is specified, that is, a state of the fuzzy set is described; the working meaning is not unique and comes before specifying a measure. The process of designing a measure [19], can conduct (as it typically happens in the applications of fuzzy set theory) to only considering P through the 'membership function' $y = m_P(x)$, and, then, to the possibility of forgetting its qualitative meaning and just considering its working meaning.

It should be clear by now, that membership functions are (ideally) measures of words that are measurable in a universe of discourse; measures designed accordingly with what at each case is available on the relation $<_P$. It is said 'ideally', since in the design process some deviation from a measure can appear. It is something similar with stating that the probability of obtaining 'five points' in throwing a single die is 1/6, by presuming the die is perfectly constructed, and that the landing surface is perfectly smooth. Is an 'ideal die'.

Membership functions should be seen as 'designed approximations' to measures, a respect at which it lacks a definition that, for instance, can be the following: A designed membership function μ_P can be seen as a 'good enough' one provided for each $\varepsilon > 0$ it exists a measure m_P such that $Im_P (x - \mu_P (x)I < \varepsilon$, for all x in X. Obviously, when μ_P is itself a measure this definition is immediately satisfied.

Measures/membership functions/states of the fuzzy set, are always designed on the basis of the current knowledge of the relation 'less P than' available to the designer, and only if the linguistic label is precise can it be potentially considered as perfectly known. Nevertheless, it does not mean that the membership function's values of an either precise, or imprecise label can be easily computed; it suffices to think in the membership function of the precise linguistic label 'transcendental' for real numbers. It is very difficult to know if a somehow defined real number is, or is not, expressible by an integer followed by a denumerable number of digits without any pattern's periodicity. It is the case for proving that the numbers π, e, e^π, the Euler-Mascheroni constant γ, the Liouville constant $\Sigma 10^{-n!}$, the Dottie number x (such that $\cos x = x$), the Chaitin halting constant Ω, etc., are actually non-algebraic, or transcendental. Proving that these numbers are transcendental requires sophisticated mathematical methods; the set of transcendental numbers has the power of continuum, is one-to-one and adjacent with all the real numbers of which they are a particular case. As Mathematics demonstrates, there are extremely complex precise concepts; precise is not a synonym of easy.

Notice that if relation $<_P$ is (artificially) identified with its sub-relation $=_P$, it means a forced 'precisification' of P in X, a change of its qualitative meaning that will imply just considering a {0, 1}—valued membership function for P, and avoiding the multiple measures that can exist for the graph $(X, <_P)$. It is a risky change of the meaning of P in X, since it can imply an understanding of P in a different form than that in current use. It can happen, for instance, if 'x is big' is understood in [0, 10] and in plain language, as 'x > 8'.

To summarize what has been said: Given a measurable linguistic label P in a universe of discourse X, its qualitative meaning generates in X a unique linguistic-collective, or fuzzy set, labeled P, that can be denoted by **P**. To consider **P** in a form allowing for its scientific and practical management, a measure of its qualitative meaning/linguistic-collective/fuzzy set, should be carefully designed to represent its contextual informational state. Once anyone of such possible states/membership functions m_P is 'designed', thanks to the contextual information furnished by $<_P$ plus the additional information (or the reasonable hypotheses) the designer will be able to add (for example, that the measure is a linear function, a triangular one, a bell shaped one, etc.), the particular use of P in X, or the linguistic-collective/fuzzy set, is just 'seen' from such membership function m_P, that is, from the currently known state of the fuzzy set. There should be an awareness of the danger that can exist when forcing an imprecise P up to be precise.

Hence, there will be cases in which the qualitative meaning $<_P$ will not fully coincide with the corresponding state's working meaning $\leq_{mP}$; the second will actually enlarge the first, and, in addition, the working meaning is not only linear but can introduce into the graph $(X, <_P)$ new arcs, that depending on the character of what is presumed for the design of the measure, can be spurious. Moreover, there can appear new working prototypes, or anti-prototypes than those that were initially recognized as such; that is, elements with a measure of one or zero that are not properly qualitative prototypes, or, respectively, qualitative anti-prototypes under $<_P$. Only seeing the meaning of P in X by a working meaning of it, implies some sort of risk.

It should still be pointed out that the usual identification of m_P (x) with a truth degree of the statement 'x is P', can result in identifying meaning with truth; that is, to simplify the concept of meaning to that of truth. Truth (= T) is a concept that, in its turn, has a meaning in the universe X [P] of the statements x is P, x $\in$ X, and that should be formerly recognized for specifying a measure m_T for the meaning of T in X [P]. Actually, such identification means accepting m_P (x) = m_T (x is P), something very risky if no qualitative meaning of T is known, and that previously requires proving that m_T (x is P) is, for all x in X, a measure of the meaning of P in X. Surely, it will require establishing some criterion of 'compatibility' [17] between P in X, and T in X [P].

4   Towards Approaching Qualitative and Working Meanings

4.1. When the working meaning coincides with the qualitative meaning, it can be said that the graph (X, $\leq_{mP}$) perfectly reproduces the initially 'observed' graph (X, $<_P$); the measure is not adding a, may be spurious, information on the use of P in X. Because of the non-linear character of $<_P$, and the linear of $\leq_{mP}$, a perfect reproduction of the previously observed qualitative meaning is not always reached; the act of designing a measure valued in the unit interval can modify what was formerly observed.

Fuzzy sets can be practically and computationally managed thanks to their states, and seeing a fuzzy set through a current state can mean a true modification of the linguistic-collective produced by changing (X, $<_P$) by (X, $\leq_{mP}$). To 'observe' a fuzzy set is only possible under a 'microscope' showing its information's states as best as possible.

Measuring can modify the initial qualitative meaning, and, obviously, the same can happen if [0, 1] were changed by whatsoever closed interval [a, b] in the real line, with prototypes taking the measure b, and anti-prototypes the measure a, and preserving property (1) of the measure.

Anyway, such a topic can be considered from two points of view. The first, is due to the practical fact that there are actual cases in which it is very difficult, if not impossible, to appreciate at each point that the measure is exactly some number in [0, 1]; for instance, sometimes it can be only recognized that such number belongs to some interval. The second, is that in plain language there are often proffered statements as, for example, 'It is highly possible that he is rich', or 'It is barely possible that she is a gifted girl', etc., whose (exact) numerical degree is surrounded by such an amount of uncertainty that it seems to better correspond to a blurred, approximate, number like 'around 0.8', 'less than 0.5', 'between 0.4 and 0, 6', 'high', etc.

Thus, there are actual situations in which it can be more suitable to adopt a range for the measure's values different from the real line, and not being linearly ordered. It is similar, for instance, to what happens when measuring the 'electrical impedance' by a complex number whose real part is the 'resistance', and its imaginary part is the 'reactance'. Recall that complex numbers are not linearly, but partially ordered, and that if its use was required by the two-fold physical phenomenon, often enough linguistic phenomena are very complex. Hence, reconsidering the range of values a measure can take cannot appear as something bizarre.

4.2. Instead of the unit interval for the values of the measure, it can be supposed a partially ordered set $(V, \leq)$, with maximum ω and minimum α. With it, the laws of a measure m ranging in V (m: $X \rightarrow V$) can be easily changed by preserving its first property, but placing ω instead of 1 in the first, and α instead of 0 in the second. With this change, the working meaning $\leq_{mP}$ is not more linear, but a partial order, whose coincidence with $<_P$ cannot be guaranteed, but it can be expected that more possibilities for it may appear. Preserved the inclusion $<_P \subseteq \leq_{mP}$, the difference set $\leq_{mP} - <_P$ have more chances to be either empty, or, at least, reduced to contain less arcs; in sum, to approach the first to the second relation. This depends on the particular problem, but it seems clear that the chances for reaching a perfect representation of the qualitative meaning by the working one will not decrease.

Two possibilities for V are the complex unit interval {a + ib; a, b $\in$ [0, 1]}, and the set of the closed sub-intervals [a, b] $\subseteq$ [0, 1], that actually are not mathematically different sets since they can be seen as isomorphic [20, 33]. Anyway, the first can have the advantage of admitting the writing of complex (Cartesian) numbers a + ib by its Euler's modulo-argument expression $\rho \ e^{i\theta}$, with $\rho = \sqrt{(a^2 + b^2)}$, and $\theta = \tan^{-1}$ (b/a) that eventually can allow geometrical considerations to be added in a given problem.

Another candidate is the set of 'fuzzy numbers', those functions [0, 1] $\rightarrow$ [0, 1] specifying linguistic labels like 'high', 'around 0.6', 'bigger than 0.4', 'between 0.3 and 0.4', etc. It is clear that the subintervals [a, b] of [0, 1] can be seen as a particular type of fuzzy numbers; for instance, the interval [0.3, 0.4] is the same as the membership function equal to 1 in it, and to zero in [0, 0.3) U (0.4, 1]. The label 'high', for example, can be represented by the identity function, or by its square, etc.

Membership functions, and in particular those of fuzzy numbers in [0, 1], are almost always point-wise ordered by,

$$\mu \leq \sigma \Leftrightarrow \mu(x) \leq \sigma(x), \text{ for all x in } [0, 1],$$

a partial order that comprises the linear ordering of crisp numbers. Its minimum is the function $\mu_0 (x) = 0$, and its maximum is $\mu_1 (x) = 1$, both for all x in X. Of course, and even if it can go against simplicity, other orderings can be chosen for the set $[0, 1]^{[0, 1]}$ (containing, at least, all membership functions) and perhaps that can be better related with a given problem. In any case, when restricted to precise numbers, such orderings should coincide with the linear order of the unit interval, and before deciding to change the unit interval by the set of 'fuzzy numbers', the designer should try to refine and improve the design.

In this way a fuzzy set P, labeled P, can be newly specified by the membership function of a fuzzy number,

$$m_P(x) = m_{\text{fuzzy number}};$$

this is, in essence, what was initially called a 'type-two fuzzy set' in the 3-part article from 1975 "The concept of a linguistic variable and its application to approximate reasoning" [33]. In the first part Zadeh introduced these type-2 fuzzy sets as follows: "... suppose that A is a fuzzy subset of a universe of discourse U, and the values of the membership function, μ_A, of A are allowed to be fuzzy subsets of the interval [0, 1]. To differentiate such fuzzy sets from those considered previously, we shall refer to them as fuzzy subsets of type 2 with the fuzzy sets whose membership functions are mappings from U to [0, 1] classified as type 1."

That is, the measure does not take numerical values at each point, but membership function ones; it is another and wider representation of the information's state of the fuzzy set, and requires to previously fix a partial ordering for the membership functions of fuzzy numbers for guaranteeing its measure's character. Since the fuzzy set is but a different name for the linguistic-collective, a 'type-two fuzzy set' refers to nothing else than a new representation of the states of the fuzzy set/linguistic-collective by means of functional values; that is, a particular type of non exclusively numerical values able to take into account the uncertainty associated to the difficulties for establishing a crisp number as a measure when it exists. It is a way of 'fuzzifying' the membership idea of 'being in' a fuzzy set, and for trying to approach the qualitative and the working meanings. More generally, in the same paper Zadeh defined then "A fuzzy set is of type n, n = 2, 3, ..., if its membership function ranges over fuzzy sets of type n-1. The membership function of a fuzzy set of type 1 ranges over the interval [0, 1]" [33].

4.3. It is still possible to take V as a set of pure words, defining the partial ordering between them, in a form that can be associated with what they mean.

A particular and suggestive case is the following. If $<_P$ is a preorder (that is, a transitive relation in addition to its presumed reflexivity), with a maximum r, and a minimum t, then the relation $=_P$ is an equivalence generating the quotient set $X/=_P$, constituted by the equivalence classes $[x] = \{y \in X; y =_P x \Longleftrightarrow x <_P y \,\&\, y <_P x\}$. These classes inherit the order of X through the definition $[x] <^* [z] \Longleftrightarrow x <_P z$; $<^*$ is a partial order, and $(X/=_P, <^*)$ is a partially ordered set with maximum $[r] = \{r\}$, and minimum $[t] = \{t\}$.

Then, the mapping $m^*: X \to X/=_P$, given by $m^*(x) = [x]$, verifies the three laws of a measure valued in the partially ordered set $(X/=_P, <^*)$, and, obviously, $<^*_{m^*}$ is isomorphic with $<_P$, that is, the working meaning perfectly reflects the qualitative meaning. Measure m^* can be named the 'natural measure'.

Provided each class [x] can be named by a word synthesizing what it represents, $(X/=_P, <^*)$ is isomorphic to the set of these words once it inherits the partial order $<^*$. Hence, and without going outside the problem's data, an example is obtained of a natural and perfect representation by words of the qualitative meaning.

4.4. There is again another view facilitated by the so-called 'interval type-two fuzzy sets' [11], employed in some applications and consisting in not considering the subintervals by their crisp membership functions, but by a kind of blurred triangular membership function for them, and more or less inspired in [33]. In this case, capturing the way of ordering these blurred functions, allowing to look at them as meaning's measures, is still difficult. It is not even clear enough what happens with prototypes and anti-prototypes; that is, how their 'measure values' are defined, what are the 'subintervals' null and unity, and if they are respectively the minimum and the maximum among them; at the end, and when linguistically describing some system, precise words can also appear in between the imprecise ones, and the maximum and minimum values preserved for them.

Mendel's 'interval type-two fuzzy sets' could be a different form of representing words, but its relation with their meaning and its measuring needs to be clarified. At least it should remain that the words considered for computing with words should be, in some clear sense, measurable.

5 Conclusion

5.1. For what has been said, expressions like 'type-two fuzzy set' are not properly appropriate since, after recognizing a qualitative meaning for the corresponding linguistic label, the fuzzy set is, in a given context, a unique, although nebulous, well anchored entity in language, or, if it is preferred, in thought. It is not to be forgotten that sets are also entities of thought; comparatively few sets that are of interest in science can be imagined like apples in a basket. As if somebody could 'see' the transcendental points between two of them in a straight line? Like sets, fuzzy sets are a creation of thoughts; like sets, fuzzy sets need some representational methodology for their practical management; for instance, that given for their membership functions/states.

Without previously knowing a qualitative meaning, it does not seem possible to follow a study allowing, in a form useful for computing with words, to practically manage the big amount of imprecise words permeating plain languages; these words should be measurable, not meaningless. Qualitative meaning, or fuzzy set, is what can be measured.

Nevertheless, to change the values for measuring the meaning's extent from the real line into a different but partially ordered structure, is something that can eventually even conduct to approaching the working and the qualitative meanings. Anyway, the character of any kind of membership functions approaching 'a measure of meaning' should be preserved.

Such change, for instance, can be particularly made within the set of fuzzy numbers in [0, 1], instead of employing the unit interval, and once fuzzy numbers are represented by some functions in $[0, 1]^{[0, 1]}$, previously endowed with a partial ordering preserving the linear one formerly existing between the 'crisp' numbers included in it. Provided such ordering is not the usual point-wise, its definition can constitute an added problem. When there is great uncertainty about the precise values of the membership function, it can be a good option, provided the ordering is defined and it has a minimum and a maximum. The problem is even more difficult if intervals are considered through a blurred kind of a triangular membership functions whose ordered structure is not well established; it waits to be clarified.

5.2. In the light of what has been presented, instead of expressions like 'type-two fuzzy set', it can be better expressed as

'fuzzy set with type-two membership function',

or something similar. In any case, it can be suitable any name not introducing in the darkness the unique, and previously existing, fuzzy set/linguistic-collective which is to be managed by means of its states, this time given by the designed membership functions of those particular fuzzy sets in [0, 1] whose linguistic label denotes either an interval, or a blurred number, instead of a crisp number. There are neither 'type-one', nor 'type-two' fuzzy sets; there are only fuzzy sets that are a purely linguistic concept.

Once X and P are given, and the graph $(X, <_P)$ is known, the linguistic-collective P is commonly and empirically recognized to exist in X. Of course, for counting with an axiomatic theory of fuzzy sets, it lacks a definition of what it can mean that two linguistic-collectives P and Q coincide in X; something that, perhaps, could be achieved by defining that their respective qualitative meanings $(X, <_P)$ and $(X, <_Q)$ are isomorphic. A different topic is how the states should be represented for practically and computationally managing the fuzzy set in a form that can be suitable for a given problem, and as it is the case, for example, when it only can be asserted that $m_P(x)$ belongs to an interval $[a(x), b(x)]$, depending on x, or that the difference set $\leq_{mP} - <_P$ is too wide.

What has been said should only be understood as a theoretical prevention against using names like 'type-two fuzzy sets', or even 'intuitionistic fuzzy sets' [1]; names that can conduct to presume the existence of different types of linguistic-collectives. But, if such types actually existed, the differences among them arise from how its qualitative meaning is expressed, but not from how the informational states can be represented at each context in a given universe of discourse. In any case, and as simplicity is of utmost importance, advising the designer: 'Never change the unit interval before trying to improve the membership function values', does not seem to be a bad advice.

In one way or another, those names could be scientifically accepted as a shortening, but shortenings should be always explained!

References

1. Atanasov, KT (2012) On intuitionistic fuzzy sets. Springer
2. Bellman RE, Kalaba RE, Zadeh LA, Abstraction and pattern classification. Memorandum RM-4307-PR. The RAND Corporation, Santa Monica, California, October 1964

3. Bellman RE, Kalaba RE, Zadeh L (1966) Abstraction and pattern classification. J Math Anal Appl 13:1–7
4. Halmos PR (1960) Naïve Set Theory. Van Nostrand, Princeton
5. Lakoff G (1973) Hedges: a study in meaning criteria and the logic of fuzzy concepts. J Philoso Log 2:458–508
6. Lakoff G (2002) Interview (Rudolf Seising), University of California at Berkeley, Dwinell Hall, 6 August 2002
7. Lee ET, Zadeh LA (1969) Note on fuzzy languages. Inf Sci 1:421–434
8. Leibniz GW (1951) Selections. Ed and trans Philip P. Wiener. Scribner's, New York
9. Leibniz GW (1923) Dissertatio de arte combinatoria, 1666, Sämtliche Schriften und Briefe, Akademie Verlag, Berlin, A VI 1, p 163; Philosophische Schriften (Gerhardt) Bd. IV
10. Ars generalis ultima. Base de dades Ramon Llull (en català). Barcelona: Centre de documentació Ramon Llull. Universitat de Barcelona, 1305
11. Mendel JM (2007) Advances in type-2 fuzzy sets and systems. Inf Sci 177:84–110
12. Rosch (1973) Natural categories. Cogn Psychol, 4:328—350
13. Sugeno M (2012) Interview (Rudolf Seising). Philos Soft Comput Newsl, 4 (1):8–12. http://docs.softcomputing.es/public/NewsletterPhilosophyAndSoftComputingNumber_6.pdf
14. Sugeno M (1972) Fuzzy measure and fuzzy integral. Trans Soc Instrum Control Eng 8 (2):218–226 (Japanese)
15. Sugeno M (1974) Theory of fuzzy integrals and its applications, PhD dissertation, Tokyo Institute of Technology
16. Trillas E, Eciolaza L (2015) Fuzzy logic. Springer
17. Trillas E (2017) On the logos. A naïve view on ordinary reasoning and fuzzy logic. Forthcoming. Springer
18. Trillas E (2006) On the use of words and fuzzy sets. Inf Sci 176(11):1463–1487
19. Trillas E, Guadarrama S (2010) Fuzzy representations need a careful design. Int J Gen Syst 39 (3):329–346
20. Trillas E, Moraga C, Termini S (2016) A naïve way of looking at fuzzy sets. Fuzzy Sets Syst 292:380–395
21. Turakainen P (1968) On Stochastic languages. Inf Control 304—313
22. Zadeh LA (1962) From circuit theory to system theory. Proc IRE 50:856–865
23. Zadeh LA (1965) Fuzzy sets and systems. In: Fox J (ed) System theory. Microwave Research Institute symposium series, vol XV, Polytech. Pr. Brooklyn, New York, pp. 29–37
24. Zadeh LA (1965) Fuzzy Sets. Inf Control 8:338–353
25. Lotfi A (1969) Toward a theory of fuzzy systems. Electronic research laboratory, College of Engineering, University of California, Berkeley 94720, Report No. ERL-69-2, June 1969
26. Zadeh LA, Fuzzy Languages and Their Relation to Human and Machine Intelligence. In: Proc. Of the Conference on Man and Computer, Bordeaux, France, 1970. Memorandum M-302, 1971, Electronics Research Laboratory, University of California, Berkeley, Calif. 94720
27. Zadeh LA (1970) Fuzzy languages and their relation to human and machine intelligence. In: Proceedings of the international conference on man and computer, Bordeaux 1970. Karger, Basel, pp 13–165
28. Zadeh LA (1971) Toward fuzziness in computer systems. Fuzzy algorithms and languages. In: Boulaye GG. (ed) Architecture and design of digital computers. Dunod, Paris, pp 9–18
29. Zadeh LA (1971) Similarity relations and fuzzy orderings. Inf Sci 3:177–200
30. Zadeh LA (1971) Quantitative fuzzy semantics. Inf Sci 3:159–176
31. Zadeh LA (1972) A fuzzy-set-theoretic interpretation of linguistic hedges. J Cybern 2:4–34
32. Zadeh LA (1973) Outline of a new approach to the analysis of complex systems and decision processes. IEEE Trans Syst Man Cybern SMC-3(1), 28–44

33. Zadeh LA (1975) The concept of a linguistic variable and its application to approximate reasoning—I. Inf Sci 8, 199–249; II, Inf Sci 8, 301–357; III, Inf Sci 9, 43–80
34. Zadeh LA (1975) Fuzzy logic and approximate reasoning. Synthese 30:407–428
35. Zadeh LA (1978) PRUF—a meaning representation language for natural languages. Int J Man-Mach Stud 10:395–460

Fuzzy Random Variables *à la* Kruse & Meyer and *à la* Puri & Ralescu: Key Differences and Coincidences

María Ángeles Gil

Abstract The concept of the so-called fuzzy random variables has been introduced in the literature aiming to model random mechanisms 'producing' fuzzy values. However, the best known approaches (namely, the one by Kwakernaak-Kruse and Meyer and the one by Féron-Puri and Ralescu) have been thought to deal with two different situations and, to a great extent, with two different probabilistic and statistical targets. This contribution highlights some of the most remarkable differences and coincidences between the two approaches.

1 Introduction

Fuzzy sets were introduced by Zadeh in 1965 [18] to model classes of objects not having a precisely defined criterion of membership. In other words, a fuzzy set is a class for which elements from a referential classical set can be compatible to a lesser or greater extent with the (not necessarily well-defined) property characterizing such a class.

Since their introduction, fuzzy set and probability theories were connected, either to emphasize distinctions between the two types of underlying uncertainties, fuzziness *vs* randomness (see Zadeh [18]) or to establish new settings and concepts involving/combining both (see, for instance, Zadeh [19]).

The author is deeply indebted to the Editors of this book because of inviting her to contribute to this tribute to Professor Rudolf Kruse. I admire Kruse a lot, and as I was invited two decades ago by Roman Slowinski, Didier Dubois and Henri Prade to coauthor with him and Jörg Gebhardt a book chapter, it was a dream that came true. Since then, I have had chance to share meetings, conferences, scientific discussions, editing a special issue, and to have him invited to our university and to Asturias for different commitments. I am delighted for having now the opportunity of saying that all this has been a great privilege for me.

M. Ángeles Gil (✉)
Facultad de Ciencias, Departamento de Estadística, I.O. y D.M,
Universidad de Oviedo, Oviedo, Spain
e-mail: magil@uniovi.es

© Springer International Publishing AG 2018
S. Mostaghim et al. (eds.), *Frontiers in Computational Intelligence*,
Studies in Computational Intelligence 739,
https://doi.org/10.1007/978-3-319-67789-7_2

In line with the second purpose, in 1976 Féron [6, 7] (see also [8]) introduced the notion of fuzzy random set to model a random mechanism 'producing' fuzzy values (more concretely, fuzzy sets of a metric space). Fuzzy random sets have been formalized either as random elements taking on values on spaces of fuzzy sets endowed with certain Borel σ-fields (i.e., following Fréchet's theory of random elements [9]) or, alternatively, as extending levelwise the notion of random sets. Féron's ideas were deeply considered and strengthened by Puri and Ralescu [15, 16] who re-coined fuzzy random sets as fuzzy random variables. Puri and Ralescu considered the specific metrics suggested by Fréchet and missing in Féron's papers, and they introduced key notions like expectation, conditional expectation, etc. In recent papers these random elements have been referred to as random fuzzy sets.

In accordance with the distinction epistemic/ontic of fuzzy values (see Couso and Dubois [4] for a recent review) Féron-Puri and Ralescu's concept corresponds to the ontic approach. Thus, fuzzy random variables in Féron-Puri and Ralescu's sense directly produce fuzzy-valued data. In fact, they are appropriate to model intrinsically imprecise-valued random attributes, like most of graded valuations associated with human ratings.

Almost simultaneously, in 1978 Kwakernaak [13, 14] introduced the notion of fuzzy random variable to formalize the fuzzy perception of an underlying real-valued random variable (called the original). In accordance with the distinction epistemic/ontic of fuzzy values, Kwakernaak's concept corresponds to the epistemic approach. Therefore, although the random mechanism behind fuzzy random variables in Kwakernaak's sense produce real-valued data, they cannot be/have not been exactly perceived but only a fuzzy perception of these data is available. Kwakernaak's ideas were formalized in a clearer mathematical way by Kruse [11] and Kruse and Meyer [12]. In summary, fuzzy random variables in Kwakernaak-Kruse-Meyer's sense are appropriate to model real-valued random attributes from which the available information is imprecise or is imprecisely reported.

This paper aims to review the key differences and analogies between these two approaches by recalling: their modelling, the way they address the formalization of the distribution and independence of fuzzy random variables, their main location and dispersion parameters, as well as a few comments about the statistical methods to analyze data from them.

2   Two Approaches to Model Fuzzy Random Variables

In this section, the definitions for the two main approaches to random mechanisms producing fuzzy values are recalled. Both definitions have been stated in a probabilistic setting in which $(\Omega, \mathscr{A}, P)$ is the probability space modelling a random experiment, where Ω is the set of all possible experimental outcomes, $\mathscr{A}$ is a σ-field of subsets of Ω (the set of all possible events of interest), and P is a probability measure associated with $(\Omega, \mathscr{A})$.

Firstly, the revisitation of Kwakernaak's conceptualization by Kruse [11] and Kruse and Meyer [12] is given as follows:

Definition 1 Let $(\Omega, \mathscr{A}, P)$ be a probability space modelling a random experiment. Let $\mathscr{F}_c(\mathbb{R})$ be the space of all fuzzy numbers (i.e., piecewise continuous, normal and fuzzy convex sets of $\mathbb{R}$). A mapping $\mathscr{X} : \Omega \to \mathscr{F}_c(\mathbb{R})$ is said to be a **fuzzy random variable *à la* Kruse & Meyer** associated with $(\Omega, \mathscr{A}, P)$ if it satisfies for each $\alpha \in (0, 1]$ that both inf $\mathscr{X}_\alpha : \Omega \to \mathbb{R}$ and sup $\mathscr{X}_\alpha : \Omega \to \mathbb{R}$ are real-valued random variables, where $\mathscr{X}_\alpha$ is the interval-valued α-level mapping, $\mathscr{X}_\alpha(\omega) = (\mathscr{X}(\omega))_\alpha = \{x \in \mathbb{R} : \mathscr{X}(\omega)(x) \geq \alpha\}$, and inf $\mathscr{X}_\alpha(\omega)$, sup $\mathscr{X}_\alpha(\omega) \in \mathscr{X}_\alpha(\omega)$, for all $\omega \in \Omega$.

Remark 1 Although it is not explicitly specified in the definition, Kwakernaak and Kruse and Meyer have clearly stated that a fuzzy random variable $\mathscr{X} : \Omega \to \mathscr{F}_c(\mathbb{R})$ in their sense is assumed to come from the composition of a real-valued random variable (the underlying one, referred to as the 'original', which is a mapping from Ω to $\mathbb{R}$) and a fuzzy perception (a mapping from $\mathbb{R}$ to $\mathscr{F}_c(\mathbb{R})$), that is,

$$\Omega \xrightarrow{\text{original random variable}} \mathbb{R} \xrightarrow{\text{fuzzy perception}} \mathscr{F}_c(\mathbb{R})$$

$$\omega \mapsto [\text{fuzzy perception} \circ \text{original random variable}](\omega) = \mathscr{X}(\omega).$$

Secondly, the revisition of Féron's conceptualization by Puri and Ralescu [15, 16] is given as follows:

Definition 2 Let $(\Omega, \mathscr{A}, P)$ be a probability space modelling a random experiment. Let $\mathscr{F}(\mathbb{R}^p)$ be the class of fuzzy subsets $\widetilde{U} : \mathbb{R}^p \to [0, 1]$ such that $\widetilde{U}_\alpha = \{x \in \mathbb{R}^p : \widetilde{U}(x) \geq \alpha\}$ is compact for each $\alpha \in (0, 1]$ and $\widetilde{U}_1 \neq \emptyset$. A **fuzzy random variable *à la* Puri & Ralescu** associated with $(\Omega, \mathscr{A}, P)$ is a mapping $\mathscr{X} : \Omega \to \mathscr{F}(\mathbb{R}^p)$ such that for each $\alpha \in (0, 1]$ the set-valued α-level mapping $\mathscr{X}_\alpha$, with $\mathscr{X}_\alpha(\omega) = (\mathscr{X}(\omega))_\alpha$ for all $\omega \in \Omega$, is a random compact set (that is, a Borel-measurable mapping with respect to the Borel σ-field generated by the topology associated with the Haussdorf metric on the space of nonempty compact subsets of $\mathbb{R}^p$).

Remark 2 It should be emphasized that Colubi et al. [2, 3] have shown that fuzzy random variables *à la* Puri & Ralescu are $\mathscr{F}(\mathbb{R}^p)$-valued random elements in Fréchet's sense, that is, they are Borel-measurable mappings w.r.t. the Borel σ-field generated by the topology associated with the Skorohod metric on $\mathscr{F}(\mathbb{R}^p)$. This Borel-measurability will be decisive for the ideas to be exposed in the next sections.

Remark 3 Although it is not explicitly specified in the definition, Féron and Puri and Ralescu have clearly stated that a fuzzy random variable $\mathscr{X} : \Omega \to \mathscr{F}_c(\mathbb{R})$ in their sense is assumed to come from the direct assessment of a fuzzy value to each experimental outcome, that is,

$$\Omega \xrightarrow{\text{fuzzy-valued random element}} \mathcal{F}_c(\mathbb{R})$$

$$\omega \mapsto [\text{fuzzy-valued random element}](\omega) = \mathcal{X}(\omega).$$

As a conclusion from Remarks 1 and 3, *the essential difference between the two approaches in defining fuzzy random variables* (appart from the dimension of the Euclidean space fuzzy values are supposed to be defined on, and the fuzzy convexity of these values) lies in the situations they model. So, fuzzy random variables *à la* Kruse & Meyer assume the existence of a real-valued random process which is fuzzily perceived (epistemic view), whereas fuzzy random variables *à la* Puri & Ralescu assume the existence of an imprecisely-valued random process for which imprecision is formalized in terms of fuzzy values (ontic view).

For some theoretical and most of practical developements, Definition 2 is particularized to the one-dimensional case ($p = 1$) and also to the fuzzy convex case (i.e., α-levels are assumed to be convex sets). Under these particularizations the two definitions match, so that *the essential coincidence between the two approaches in defining fuzzy random variables* can be stated (see, for instance, Blanco-Fernández et al. [1] for a recent review about) as follows:

Proposition 1 *Let* $(\Omega, \mathcal{A}, P)$ *be a probability space modelling a random experiment and let* $\mathcal{F}_c(\mathbb{R})$ *be the space of all fuzzy numbers. Then, a mapping* $\mathcal{X} : \Omega \to \mathcal{F}_c(\mathbb{R})$ *is a fuzzy random variable* à la Kruse & Meyer *if, and only if, it is a fuzzy random variable* à la Puri & Ralescu.

3 Distribution and Independence of Fuzzy Random Variables

As we have remarked in the preceding sections, the situations the two approaches to fuzzy random variables have been motivated on are different. This fact is crucial to support a key concept in dealing with probabilistic and statistical developments involving fuzzy random variables: the distribution and the independence.

If $\mathcal{X}$ is a fuzzy random variable *à la* Kruse & Meyer, its distribution is considered to be propagated from the distribution of the original through the fuzzy perception. This propagation is carried out on the basis of Zadeh's extension principle [20], so that the (fuzzy-valued) **distribution function of fuzzy random variable** $\mathcal{X}$ associated with $(\Omega, \mathcal{A}, P)$ is defined as follows:

$$\widetilde{F}_{\mathscr{X}} : \mathbb{R} \to \mathscr{F}_c(\mathbb{R}), \quad x \mapsto \widetilde{F}_{\mathscr{X}}(x) : \mathbb{R} \to [0, 1]$$

$$\left(\widetilde{F}_{\mathscr{X}}(x)\right)(p) = \begin{cases} \sup\limits_{X_0 \in \mathrm{Orig}(\mathscr{X}) \,:\, F_{X_0}(x)=p} \inf\limits_{\omega \in \Omega} \mathscr{X}(\omega)(X_0(\omega)) & \text{if } p \in [0, 1] \\ 0 & \text{otherwise} \end{cases}$$

where $\mathrm{Orig}(\mathscr{X})$ is the set of potential originals of $\mathscr{X}$ and F_X denotes the distribution function of random variable X associated with $(\Omega, \mathscr{A}, P)$.

As a consequence from this notion, two fuzzy random variables $\mathscr{X}$ and $\mathscr{Y}$ *à la* Kruse & Meyer are defined to be **identically distributed fuzzy random variables** if their fuzzy distribution functions $\widetilde{F}_{\mathscr{X}}$ and $\widetilde{F}_{\mathscr{Y}}$ coincide for all $x \in \mathbb{R}$. This has been shown (see Kruse and Meyer [12]) to be equivalent to say that $\mathscr{X}$ and $\mathscr{Y}$ are identically distributed fuzzy random variables if and only if for each $\alpha \in (0, 1]$ the random variables $\inf \mathscr{X}_\alpha$ and $\inf \mathscr{Y}_\alpha$ are identically distributed and $\sup \mathscr{X}_\alpha$ and $\sup \mathscr{Y}_\alpha$ are also identically distributed.

In a similar way, fuzzy random variables $\mathscr{X}_1, \ldots, \mathscr{X}_n$ *à la* Kruse & Meyer are defined to be **(either pairwise or completely) independent fuzzy random variables** if their joint fuzzy distribution functions can be (pairwise or completely) factorized in terms of the marginals. This has been shown (see Kruse and Meyer [12]) to be equivalent to say that $\mathscr{X}_1, \ldots, \mathscr{X}_n$ are (pairwise or completely) independent fuzzy random variables if and only if for each $\alpha \in (0, 1]$ random variables $\inf(\mathscr{X}_1)_\alpha, \ldots,$ $\inf(\mathscr{X}_n)_\alpha$ are (respectively, pairwise or completely) independent and $\sup(\mathscr{X}_1)_\alpha, \ldots,$ $\sup(\mathscr{X}_n)_\alpha$ are also (respectively, pairwise or completely) independent.

On the other hand, if $\mathscr{X}$ is a fuzzy random variable *à la* Puri & Ralescu, its distribution function cannot be directly extended, due to the lack of a universally accepted total order on the space of fuzzy values. However, because of the Borel-measurability of a fuzzy random variable *à la* Puri & Ralescu (Remark 2), one can immediately induce the distribution of this random element from the probability measure P in $(\Omega, \mathscr{A}, P)$, so that for any Borel set $\mathfrak{B}$ of fuzzy values in $\mathscr{F}(\mathbb{R}^p)$, the (real-valued) **induced probability of $\mathfrak{B}$ by** $\mathscr{X}$ is given (without need to be specifically defined) by

$$P(\mathscr{X} \in \mathfrak{B}) = P(\{\omega \in \Omega \,:\, \mathscr{X}(\omega) \in \mathfrak{B}\}).$$

Analogously, the notions of identity in distribution and independence can be immediately derived on the basis of the Borel-measurability assumption for fuzzy random variables *à la* Puri & Ralescu. Therefore, without need to be specifically defined, so that two fuzzy random variables $\mathscr{X}$ and $\mathscr{Y}$ *à la* Puri & Ralescu are **identically distributed fuzzy random variables** if and only if for any Borel set $\mathfrak{B}$ of fuzzy values $P(\mathscr{X} \in \mathfrak{B}) = P(\mathscr{Y} \in \mathfrak{B})$.

And, fuzzy random variables $\mathscr{X}_1, \ldots, \mathscr{X}_n$ *à la* Puri & Ralescu are **(either pairwise or completely) independent fuzzy random variables** if for Borel sets $\mathfrak{B}_1, \ldots,$ $\mathfrak{B}_n$ of fuzzy values

$$P(\mathscr{X}_i \in \mathfrak{B}_i, \mathscr{X}_j \in \mathfrak{B}_j) = P(\mathscr{X}_i \in \mathfrak{B}_i) \cdot P(\mathscr{X}_j \in \mathfrak{B}_j) \quad \text{for } i \neq j,$$
$$P(\mathscr{X}_1 \in \mathfrak{B}_1, \ldots, \mathscr{X}_n \in \mathfrak{B}_n) = P(\mathscr{X}_1 \in \mathfrak{B}_1) \cdot \ldots \cdot P(\mathscr{X}_n \in \mathfrak{B}_n).$$

Although the way to formalize the distribution of fuzzy random variables depends on the considered approach, so there is an essential difference in managing the distribution, the identity in distribution and independence in Puri & Ralescu's approach imply those in Kruse & Meyer's one in case one deals with fuzzy number-valued random variables.

4 Parameters of the Distribution of Fuzzy Random Variables

In summarizing the distribution of fuzzy random variables, the most used summary measures are location and dispersion ones. To extend them from the real-valued case the way to proceed depends on the approach and also on the way the distribution has been formalized.

In connection with the distribution of fuzzy random variables *à la* Kruse & Meyer, extension is based on Zadeh's extension principle, so that a fuzzy parameter of a fuzzy random variable $\mathscr{X}$ is viewed as a fuzzy perception of a real-valued parameter of the original. Thus, if $\theta(X)$ denotes the parameter of the original X the extended **fuzzy perception of** θ for $\mathscr{X}$ is given by the fuzzy number $\widetilde{\vartheta}(\mathscr{X})$ such that for each $t \in \mathbb{R}$

$$\widetilde{\vartheta}(\mathscr{X})(t) = \sup_{X_0 \in \mathrm{Orig}(\mathscr{X})\,:\,\theta(X_0)=t} \inf_{\omega \in \Omega} \mathscr{X}(\omega)(X_0(\omega)).$$

In accordance with this extension, if $\mathscr{X}$ is an *à la* Kruse & Meyer fuzzy random variable the main location measures are given by

- the **fuzzy perception of the mean** corresponds to

$$\widetilde{E}(\mathscr{X})(t) = \sup_{X_0 \in \mathrm{Orig}(\mathscr{X})\,:\,E(X_0)=t} \inf_{\omega \in \Omega} \mathscr{X}(\omega)(X_0(\omega)),$$

which satisfies for each $\alpha \in (0, 1]$ (see Kruse [11]) that

$$\left(\widetilde{E}(\mathscr{X})\right)_\alpha = \left[E(\inf \mathscr{X}_\alpha), E(\sup \mathscr{X}_\alpha)\right];$$

- the **fuzzy perception of the median** corresponds to

$$\widetilde{\Gamma}(\mathscr{X})(t) = \sup_{X_0 \in \mathrm{Orig}(\mathscr{X})\,:\,\mathrm{Me}(X_0)=t} \inf_{\omega \in \Omega} \mathscr{X}(\omega)(X_0(\omega)),$$

which satisfies that for each $\alpha \in (0, 1]$ (see Grzegorzewski [10])

$$\left(\widetilde{\Gamma}(\mathscr{X})\right)_\alpha = [\underline{\mathrm{Me}}(\inf \mathscr{X}_\alpha), \overline{\mathrm{Me}}(\sup \mathscr{X}_\alpha)],$$

where $\underline{Me}/\overline{Me}$ denotes the median of the corresponding real-valued random variable with the convention (if the median is not unique) of taking the smallest/largest median.

In connection with the distribution of fuzzy random variables *à la* Puri & Ralescu, extension is based on Fréchet's ideas [9] for random elements over metric spaces, so that if one considers the L^p metrics ($p \in \{1, 2\}$) by Diamond and Kloeden [5] on $\mathcal{F}_c(\mathbb{R})$

$$\rho_p(\widetilde{U}, \widetilde{V}) = \left[\frac{1}{2} \int_{(0,1]} \left(|\inf \widetilde{U}_\alpha - \inf \widetilde{V}_\alpha|^p + |\sup \widetilde{U}_\alpha - \sup \widetilde{V}_\alpha|^p \right) d\alpha \right]^{1/p}.$$

In accordance with Fréchet's ideas, if $\mathcal{X}$ is an *à la* Puri & Ralescu fuzzy random variable

• the **extended mean** corresponds to

$$\widetilde{E}(\mathcal{X}) = \arg \min_{\widetilde{U} \in \mathcal{F}(\mathbb{R}^p)} E\left(\left[\rho_2(\mathcal{X}, \widetilde{U}) \right]^2 \right),$$

which satisfies for each $\alpha \in (0, 1]$ (see Puri and Ralescu [16]) that $\left(\widetilde{E}(\mathcal{X}) \right)_\alpha$ = Aumann integral of $\mathcal{X}_\alpha$, and coincides with $\left[E(\inf \mathcal{X}_\alpha), E(\sup \mathcal{X}_\alpha) \right]$ if $\mathcal{X}$ is $\mathcal{F}_c(\mathbb{R})$-valued;

• if $\mathcal{X}$ is $\mathcal{F}_c(\mathbb{R})$-valued, the **extended (1-norm) median** corresponds to

$$\widetilde{Me}(\mathcal{X}) = \arg \min_{\widetilde{U} \in \mathcal{F}(\mathbb{R}^p)} E\left(\rho_1(\mathcal{X}, \widetilde{U}) \right),$$

for which a solution (see Sinova et al. [17]) is the one such that for each $\alpha \in (0, 1]$ is given by $\left(\widetilde{Me}(\mathcal{X}) \right)_\alpha = [Me(\inf \mathcal{X}_\alpha), Me(\sup \mathcal{X}_\alpha)]$, where Me denotes the median of the corresponding real-valued random variable with the convention (if the median is not unique) of taking the middle median. Actually, some other conventions could be considered to choose Me (like the one leading to $\widetilde{\Gamma}(\mathcal{X})$), whenever a fuzzy number is determined.

Regarding the **variance**, the policy for this approach is essentially different from an approach to the other. If $\mathcal{X}$ is an *à la* Kruse & Meyer fuzzy random variable and Zadeh's extension principle is applied, the variance is conceived as a fuzzy perception of the variance of the original, so it is fuzzy-valued. If $\mathcal{X}$ is an *à la* Puri & Ralescu fuzzy random variable and Fréchet's ideas are applied, the variance is conceived as a real-valued measure given by the mean squared L^2 distance between the fuzzy random variable and its mean value. Consequently, they stand for two different types of measures.

5 Statistical Data Analysis from Fuzzy Random Variables

The approach behind the two examined notions for fuzzy random variables influences the statistical data analysis one can develop. Since there is a wide class of methods to estimation and testing hypothesis from fuzzy data, the topic cannot be entered in this paper.

Anyway, it is interesting to highlight that analyses concerning fuzzy random variables *à la* Kruse & Meyer can refer either to parameters of the originals or to fuzzy perceptions of them (see, for instance, Kruse and Meyer [12] for several examples). In contrast to this, analyses concerning fuzzy random variables *à la* Puri & Ralescu always refer to parameters of the distribution of the fuzzy random variables (see for instance, Blanco-Fernández et al. [1]). Actually, in connection with the last ones, it should be pointed out that, thanks to having modelled fuzzy random variables as random elements, all the basic concepts from statistics with crisp data (e.g., unbiased estimation, p-values, and do on, etc.) can be preserved without needing to expressly define them.

Following Kruse, the analysis of fuzzy-valued data should be clearly distinguished from the analysis of data by using fuzzy logic-based methods.

Acknowledgements The research in this paper has been partially supported by the Principality of Asturias/FEDER Grant GRUPIN14-101 and the Spanish Ministry of Economy and Competitiveness Grant MTM2015-63971-P. Their financial support is gratefully acknowledged.

References

1. Blanco-Fernández A, Casals MR, Colubi A, Corral N, García-Bárzana M, Gil MA, González-Rodríguez G, López MT, Lubiano MA, Montenegro M, Ramos-Guajardo AB, de la Rosa de Sáa S, Sinova B (2014) A distance-based statistical analysis of fuzzy number-valued data. Int J Approx Reas 55:1487–1501
2. Colubi A, Domínguez-Menchero JS, López-Díaz M, Ralescu DA (2001) On the formalization of fuzzy random variables. Inform Sci 133:3–6
3. Colubi A, Domínguez-Menchero JS, López-Díaz M, Ralescu DA (2002) A $D_E[0, 1]$ representation of random upper semicontinuous functions. Proc Amer Math Soc 130:3237–3242
4. Couso I, Dubois D (2014) Statistical reasoning with set-valued information: ontic vs. epistemic views. Int J Approx Reas 55:1502–1518
5. Diamond P, Kloeden P (1990) Metric spaces of fuzzy sets. Fuzzy Sets Syst 35:241–249
6. Féron R (1976) Ensembles aléatoires flous. CR Acad Sci Paris A 282:903–906
7. Féron R (1976) Ensembles flous attachés à un ensemble aléatoire flou. Publ Econometr 9:51–66
8. Féron R (1979) Sur les notions de distance et d'ecart dans une structure floue et leurs applications aux ensembles aléatoires flous. CR Acad Sci Paris A 289:35–38
9. Fréchet M (1948) Les éléments aléatoires de nature quelconque dans un espace distancié. Ann L'Inst H Poincaré 10:215–310
10. Grzegorzewski P (1998) Statistical inference about the median from vague data. Contr Cyber 27:447–464
11. Kruse R (1982) The strong law of large numbers for fuzzy random variables. Inform Sci 28:233–241

12. Kruse R, Meyer KD (1987) Statistics with vague data. Reidel, Dordrecht
13. Kwakernaak H (1978) Fuzzy random variables, part I: definitions and theorems. Inform Sci 15:1–15
14. Kwakernaak H (1979) Fuzzy random variables, part II: algorithms and examples for the discrete case. Inform Sci 17:253–278
15. Puri ML, Ralescu DA (1985) The concept of normality for fuzzy random variables. Ann Probab 11:1373–1379
16. Puri ML, Ralescu DA (1986) Fuzzy random variables. J Math Anal Appl 114:409–422
17. Sinova B, Gil MA, Colubi A, Van Aelst S (2012) The median of a random fuzzy number. The 1-norm distance approach. Fuzzy Sets Syst 200:99–115
18. Zadeh LA (1965) Fuzzy sets. Inform Contr 8:338–353
19. Zadeh LA (1968) Probability measures of fuzzy events. J Math Anal Appl 23:421–427
20. Zadeh LA (1975) The concept of a linguistic variable and its application to approximate reasoning. part 1: Inform Sci 8:199–249; part 2: Inform Sci 8:301–353; part 3: Inform Sci 9:43–80

Statistical Inference for Incomplete Ranking Data: A Comparison of Two Likelihood-Based Estimators

Inés Couso and Eyke Hüllermeier

Abstract We consider the problem of statistical inference for ranking data, namely the problem of estimating a probability distribution on the permutation space. Since observed rankings could be incomplete in the sense of not comprising all choice alternatives, we propose to tackle the problem as one of learning from imprecise or coarse data. To this end, we associate an incomplete ranking with its set of consistent completions. We instantiate and compare two likelihood-based approaches that have been proposed in the literature for learning from set-valued data, the marginal and the so-called face-value likelihood. Concretely, we analyze a setting in which the underlying distribution is Plackett-Luce and observations are given in the form of pairwise comparisons.

1 Introduction

The study of rank data and related probabilistic models on the permutation space (symmetric group) has a long tradition in statistics, and corresponding methods for parameter estimation have been used in various fields of application, such as psychology and the social sciences [1]. More recently, applications in information retrieval (search engines) and machine learning (personalization, preference learning) have caused a renewed interest in the analysis of rankings and topics such as "learning-to-rank" [2]. Indeed, methods for learning and constructing preference models from explicit or implicit preference information and feedback are among the recent research trends in these disciplines [3].

In most applications, the rankings observed are *incomplete* or *partial* in the sense of including only a subset of the underlying choice alternatives, whereas no preferences are revealed about the remaining ones—pairwise comparisons can be seen as

I. Couso
University of Oviedo, Oviedo, Spain
e-mail: couso@uniovi.es

E. Hüllermeier (✉)
Paderborn University, Paderborn, Germany
e-mail: eyke@upb.de

© Springer International Publishing AG 2018
S. Mostaghim et al. (eds.), *Frontiers in Computational Intelligence*,
Studies in Computational Intelligence 739,
https://doi.org/10.1007/978-3-319-67789-7_3

an important special case. In this paper, we therefore approach the problem of learning from ranking data from the point of view of statistical inference with *imprecise data*. The key idea is to consider an incomplete ranking as a *set-valued observation*, namely the set of all complete rankings consistent with the incomplete observation [4]. This approach is especially motivated by recent work on learning from imprecise, incomplete, or fuzzy data [5–9].

Thus, our paper can be seen as an application of general methods proposed in that field to the specific case of ranking data. This is arguably interesting for both sides, research on statistics with imprecise data and learning from ranking data: For the former, ranking data is an interesting test bed that may help understand, analyze, and compare methods for learning from imprecise data; for the latter, general approaches for learning from imprecise data may turn into new statistical methods for ranking.

In this paper, we compare two likelihood-based approaches for learning from imprecise data. More specifically, both approaches are used for inference about the so-called Plackett-Luce model, a parametric family of probability distributions on the permutation space.

2 Preliminaries and Notation

Let $\mathbb{S}_K$ denote the collection of rankings (permutations) over a set $U = \{a_1, \ldots, a_K\}$ of K items a_k, $k \in [K] = \{1, \ldots, K\}$. A complete ranking (a generic element of $\mathbb{S}_K$) is a bijection $\pi : [K] \longrightarrow [K]$, where $\pi(k)$ is the position of the kth item a_k in the ranking. We denote by π^{-1} the ordering associated with a ranking, i.e., $\pi^{-1}(j)$ is the index of the item on position j. We write rankings in brackets and orderings in parentheses; for example, $\pi = [2, 4, 3, 1]$ and $\pi^{-1} = (4, 1, 3, 2)$ both denote the ranking $a_4 \succ a_1 \succ a_3 \succ a_2$.

For a possibly incomplete ranking, which includes only some of the items, we use the symbol τ (instead of π). If the kth item does not occur in a ranking, then $\tau(k) = 0$ by definition; otherwise, $\tau(k)$ is the rank of the kth item. In the corresponding ordering, the missing items do simply not occur. For example, the ranking $a_4 \succ a_1 \succ a_2$ would be encoded as $\tau = [2, 3, 0, 1]$ and $\tau^{-1} = (4, 1, 2)$, respectively. We let $I(\tau) = \{k \ : \ \tau(k) > 0\} \subset [K]$ and denote the set of all rankings (complete or incomplete) by $\overline{\mathbb{S}}_K$.

An incomplete ranking τ can be associated with its set of consistent extensions $E(\tau) \subset \mathbb{S}_K$, where

$$E(\tau) = \Big\{ \pi \ : \ (\tau(i) - \tau(j))(\pi(i) - \pi(j)) \geq 0 \text{ for all } i, j \in I(\tau) \Big\}$$

An important special case is an incomplete ranking $\tau_{i,j}$ in the form of a pairwise comparison $a_i \succ a_j$ (i.e., $\tau_{i,j}(i) = 1$, $\tau_{i,j}(j) = 2$, $\tau_{i,j}(k) = 0$ otherwise), which is associated with the set of extensions

$$E(\tau_{i,j}) = E(a_i \succ a_j) = \{\pi \in \mathbb{S}_K \; : \; \pi(i) < \pi(j)\} \; .$$

Modeling an incomplete observation τ by the set of linear extensions $E(\tau)$ reflects the idea that τ has been produced from an underlying complete ranking π by some "coarsening" or "imprecisiation" process, which essentially consists of omitting some of the items from the ranking. $E(\tau)$ then corresponds to the set of all possible candidates π, i.e., all complete rankings that are compatible with the observation τ if nothing is known about the coarsening, except that it does not change the relative order of any items.

Sometimes, more knowledge about the coarsening is available, or reasonable assumptions can be made. For example, it might be known that τ is a top-t ranking, which means that it consists of the items that occupy the first t positions in π.

3 Probabilistic Models

Statistical inference requires a probabilistic model of the underlying data generating process, which, in our case, essentially comes down to specifying a probability distribution on the permutation space. One of the most well-known probability distributions of that kind is the Plackett-Luce (PL) model [1].

3.1 The Plackett-Luce Model

The PL model is parametrized by a vector $\theta = (\theta_1, \theta_2, \dots, \theta_K) \in \Theta = \mathbb{R}_+^K$. Each θ_i can be interpreted as the weight or "strength" of the option a_i. The probability assigned by the PL model to a ranking represented by a permutation $\pi \in \mathbb{S}_K$ is given by

$$\mathrm{pl}_\theta(\pi) = \prod_{i=1}^{K} \frac{\theta_{\pi^{-1}(i)}}{\theta_{\pi^{-1}(i)} + \theta_{\pi^{-1}(i+1)} + \cdots + \theta_{\pi^{-1}(K)}} \tag{1}$$

Obviously, the PL model is invariant toward multiplication of θ with a constant $c > 0$, i.e., $\mathrm{pl}_\theta(\pi) = \mathrm{pl}_{c\theta}(\pi)$ for all $\pi \in \mathbb{S}_K$ and $c > 0$. Consequently, θ can be normalized without loss of generality (and the number of degrees of freedom is only $K - 1$ instead of K). Note that the most probable ranking, i.e., the mode of the PL distribution, is simply obtained by sorting the items in decreasing order of their weight:

$$\pi^* = \arg\max_{\pi \in \mathbb{S}_K} \mathrm{pl}_\theta(\pi) = \arg\operatorname*{sort}_{k \in [K]}\{\theta_1, \dots, \theta_K\} \; . \tag{2}$$

As a convenient property of PL, let us mention that it allows for a very easy computation of marginals, because the marginal probability on a subset $U' = \{a_{i_1}, \dots, a_{i_j}\} \subset$

U of $J \leq K$ items is again a PL model parametrized by $(\theta_{i_1}, \ldots, \theta_{i_J})$. Thus, for every $\tau \in \overline{\mathbb{S}}_K$ with $I(\tau) = U'$,

$$
\mathrm{pl}_\theta(\tau) = \sum_{\pi \in E(\tau)} \mathrm{pl}_\theta(\pi) = \prod_{j=1}^{J} \frac{\theta_{\tau^{-1}(j)}}{\theta_{\tau^{-1}(j)} + \theta_{\tau^{-1}(j+1)} + \cdots + \theta_{\tau^{-1}(J)}} \tag{3}
$$

In particular, this yields pairwise probabilities

$$
\mathrm{pl}_\theta(\tau_{i,j}) = \mathrm{pl}_\theta(a_i \succ a_j) = \frac{\theta_i}{\theta_i + \theta_j} .
$$

This is the well-known Bradley-Terry model [1], a model for the pairwise comparison of alternatives. Obviously, the larger θ_i in comparison to θ_j, the higher the probability that a_i is chosen. The PL model can be seen as an extension of this principle to more than two items: the larger the parameter θ_i in (1) in comparison to the parameters $\theta_j, j \neq i$, the higher the probability that a_i occupies a top rank.

3.2 A Stochastic Model of Coarsening

While (1) defines a probability for every *complete* ranking π, and hence a distribution $p : \mathbb{S}_K \longrightarrow [0, 1]$, an extension of p from $\mathbb{S}_K$ to $\overline{\mathbb{S}}_K$ is in principle offered by (3). One should note, however, that marginalization in the traditional sense is different from coarsening. In fact, (3) assumes the subset of items U' to be fixed beforehand, prior to drawing a ranking at random. For example, focusing on two items a_i and a_j, one may ask for the probability that a_i will precede a_j in the next ranking drawn at random according to p.

Recalling our idea of a coarsening process, it is more natural to consider the data generating process as a two step procedure:

$$
p_{\theta,\lambda}(\tau, \pi) = p_\theta(\pi) \cdot p_\lambda(\tau \mid \pi) \tag{4}
$$

According to this model, a complete ranking π is generated first according to $p_\theta(\cdot)$, and this ranking is then turned into an incomplete ranking τ according to $p_\lambda(\cdot \mid \pi)$. Thus, the coarsening process is specified by a family of conditional probability distributions

$$
\left\{ p_\lambda(\cdot \mid \pi) : \pi \in \mathbb{S}_K, \lambda \in \Lambda \right\} , \tag{5}
$$

where λ collects all parameters of these distributions; $p_{\theta,\lambda}(\tau, \pi)$ is the probability of producing the data $(\tau, \pi) \in \overline{\mathbb{S}}_K \times \mathbb{S}_K$. Note, however, that π is actually not observed.

4   Statistical Inference

As for the statistical inference about the process (4), our main interest concerns the "precise part", i.e., the parameter θ, whereas the coarsening is rather considered as a complication of the estimation. In other words, we are less interested in inference about λ or, stated differently, we are interested in λ only in so far as it helps to estimate θ. In this regard, it should also be noted that inference about λ will generally be difficult: Due to the sheer size of the family of distributions (5), λ could be very high-dimensional. Besides, concrete model assumptions about the coarsening process may not be obvious.

Therefore, what we are mainly aiming for is an estimation technique that is efficient in the sense of circumventing direct inference about λ, and at the same time robust in the sense that it yields reasonably good results for a wide range of coarsening procedures, i.e., under very weak assumptions about the coarsening (or perhaps no assumptions at all). As a first step toward this goal, we look at two estimation principles that have recently been proposed in the literature, both being based on the principle of likelihood maximization.

In the following, the random variable X will denote the precise outcome of a single random experiment, i.e., a complete ranking π, whereas Y denotes the coarsening τ. We assume to be given an i.i.d. sample of size N and let $\tau = (\tau_1, \dots, \tau_N) \in (\overline{\mathbb{S}}_K)^N$ denote a sequence of N independent incomplete observations of Y.

4.1   The Marginal Likelihood

The perhaps most natural approach is to consider the marginal likelihood function (also called "visible likelihood" in [10]), i.e., the probability of the observed data Y given the parameters θ and λ:

$$L_V(\theta, \lambda) = p(\tau \mid \theta, \lambda) = \prod_{i=1}^{N} p(Y = \tau_i \mid \theta, \lambda)$$

$$= \prod_{i=1}^{N} \sum_{\pi \in \mathbb{S}_K} p_\theta(\pi) p_\lambda(\tau_i \mid \pi) \qquad (6)$$

The maximum likelihood estimate (MLE) would then be given by

$$(\theta^*, \lambda^*) = \arg\max_{(\theta, \lambda) \in \Theta \times \Lambda} L_V(\theta, \lambda),$$

or, emphasizing inference about θ, by

$$\theta^* = \arg\max_{\theta \in \Theta} \max_{\lambda \in \Lambda} L_V(\theta, \lambda).$$

As can be seen, this approach requires assumptions about the parametrization of the coarsening, i.e., the parameter space Λ. Of course, since both $\mathbb{S}_K$ and $\overline{\mathbb{S}}_K$ are finite, these assumptions can be "vacuous" in the sense of allowing all possible distributions. Thus, the family (5) would be specified in a tabular form by letting

$$p_\lambda(\tau \mid \pi) = \lambda_{\pi,\tau} \tag{7}$$

for all $\tau \in \overline{\mathbb{S}}_K$ and $\pi \in E(\tau)$ (recall that $p_\lambda(\tau \mid \pi) = 0$ for $\pi \notin E(\tau)$). In other words, Λ is given by the set of all these parametrizations under the constraint that

$$\sum_{\tau \in \overline{\mathbb{S}}_K} \lambda_{\pi,\tau} = \sum_{\tau \in E(\tau)} \lambda_{\pi,\tau} = 1$$

for all $\pi \in \mathbb{S}_K$. We denote this parametrization by Λ_{vac}.

4.2   The Face-Value Likelihood

The *face-value likelihood* is expressed as follows [11, 12]:

$$L_F(\theta, \lambda) = \prod_{i=1}^N P\big(X \in E(\tau_i) \mid \theta, \lambda\big) \tag{8}$$

$$= \prod_{i=1}^N \sum_{\pi \in E(\tau_i)} p_\theta(\pi)$$

Note that the face-value likelihood does actually not depend on λ, which means that we could in principle write $L_F(\theta)$ instead of $L_F(\theta, \lambda)$. Indeed, this approach does not explicitly account for the coarsening process, or at least does not consider the coarsening as a stochastic process. The only way of incorporating knowledge about this process is to replace the set of linear extensions, $E(\tau_i)$, with a smaller set of complete rankings associated with an incomplete observation τ_i. This can be done if the coarsening is deterministic, like in the case of top-t selection.

5   Comparison of the Approaches

These two likelihood functions (6) and (8) coincide when the collection of possible values for Y forms a partition of the collection of permutations $\mathbb{S}_K$, since the events $Y = \tau_i$ and $X \in E(\tau_i)$ are then the same. But they do not coincide in the general case, where the event $Y = \tau_i$ implies but does not necessarily coincide with $X \in E(\tau_i)$.

In the following, we refer to the parameter estimation via maximization of (6) and (8) as MLM (marginal likelihood maximization) and FLM (face-value likelihood maximization), respectively.

5.1 Known Coarsening

A comparison between the marginal and face-value likelihood is arguable in the case where the coarsening is assumed to be known, because, as already said, the face-value likelihood is not able to exploit this knowledge (unless the coarsening is deterministic and forms a partition). Obviously, ignorance of the coarsening may lead to very poor estimates in general, as shown by the following example.

Let $K = 3$ and $U = \{a_1, a_2, a_3\}$. To simplify notation, we denote a ranking $a_i \succ a_j \succ a_k$ inducing $\pi^{-1} = (i, j, k)$ by $a_i a_j a_k$. We assume the PL model and suppose the coarsening to be specified by the following (deterministic) relation between complete rankings π and incomplete observations τ, which are all given in the form of pairwise comparisons:

	$a_1 \succ a_2$	$a_2 \succ a_1$	$a_1 \succ a_3$	$a_3 \succ a_1$	$a_2 \succ a_3$	$a_3 \succ a_2$
$a_1 a_2 a_3$	0	0	0	0	1	0
$a_1 a_3 a_2$	0	0	0	0	0	1
$a_2 a_1 a_3$	0	1	0	0	0	0
$a_2 a_3 a_1$	0	1	0	0	0	0
$a_3 a_1 a_2$	0	0	0	1	0	0
$a_3 a_2 a_1$	0	1	0	0	0	0

Denoting by $n_{i,j}$ the number of times $a_i \succ a_j$ has been observed, the face-value likelihood function reads as follows:

$$L_F(\tau; \theta) = \prod_{i=1}^{3} \prod_{j \neq i} \left(\frac{\theta_i}{\theta_i + \theta_j} \right)^{n_{i,j}}.$$

Let now n_{ijk} denote the number of occurrences of the ranking $a_i a_j a_k$ in the sample. According to the above relation, we have the following:

$$n_{1,2} = 0$$
$$n_{2,1} = n_{213} + n_{231} + n_{321}$$
$$n_{1,2} = 0$$
$$n_{1,3} = n_{312}$$
$$n_{2,3} = n_{123}$$
$$n_{3,2} = n_{132}$$

Therefore,

$$L_F(\theta) = \left(\frac{\theta_2}{\theta_1 + \theta_2}\right)^{n_{213} + n_{231} + n_{321}}$$

$$\times \left(\frac{\theta_3}{\theta_1 + \theta_3}\right)^{n_{312}} \times \left(\frac{\theta_2}{\theta_2 + \theta_3}\right)^{n_{123}} \times \left(\frac{\theta_3}{\theta_2 + \theta_3}\right)^{n_{132}}.$$

For an arbitrary triplet $\theta = (\theta_1, \theta_2, \theta_2)$ with $\theta_1 + \theta_2 + \theta_3 = 1$, we observe that

$$L_F(\theta_1, \theta_2, \theta_3) \le L_F\left(0, \theta_2', \theta_3'\right),$$

where $\theta_2' = \frac{\theta_2}{\theta_2 + \theta_3}$ and $\theta_3' = \frac{\theta_3}{\theta_2 + \theta_3}$. In fact,

$$L_F(0, \theta_2', \theta_3') = (\theta_2')^{n_{123}} \cdot (\theta_3')^{n_{132}},$$

and therefore

$$L_F(\theta) = \left[\left(\frac{\theta_2}{\theta_1 + \theta_2}\right)^{n_{213} + n_{231} + n_{321}} \cdot \left(\frac{\theta_3}{\theta_1 + \theta_3}\right)^{n_{312}}\right] \times L_F(0, \theta_2', \theta_3'),$$

which is clearly less than or equal to $L_F(0, \theta_2', \theta_3')$. Furthermore, according to Gibb's inequality, the above likelihood value, $L_F(0, \theta_2', \theta_3')$, is maximized for

$$(\hat{\theta}_2', \hat{\theta}_3') = \left(\frac{n_{123}}{n_{123} + n_{132}}, \frac{n_{132}}{n_{123} + n_{132}}\right).$$

For instance, if we assume that the true distribution over $\mathbb{S}_3$ is PL with parameter $\theta = (\theta_1, \theta_2, \theta_3) = (0.99, 0.005, 0.005)$, then our estimation of θ based on the face-value likelihood function will be

$$\left(0, \frac{n_{123}}{n_{123} + n_{132}}, \frac{n_{132}}{n_{123} + n_{132}}\right),$$

which tends to $(0, 0.5, 0.5)$ as n tends to infinity.

5.2 Unknown Coarsening

The comparison between the two approaches appears to be more reasonable when the coarsening is assumed to be unknown. In that case, it might be fair to instantiate the marginal likelihood with the parametrization Λ_{vac}, because just like the face-value likelihood, it is then essentially ignorant about the coarsening. However, the estimation of the coarsening process under Λ_{vac} is in general not practicable, simply because the number of parameters (7) is too large: One parameter $\lambda_{\pi,\tau}$ for each $\tau \in \bar{\mathbb{S}}_K$ and $\pi \in E(\tau)$ makes about $2^K K!$ parameters in total. Besides, Λ_{vac} may cause problems of model identifiability. What we need, therefore, is a simplifying assumption on the coarsening.

5.2.1   Rank-Dependent Coarsening

The assumption we make here is a property we call *rank-dependent* coarsening. A coarsening procedure is rank-dependent if the incompletion is only acting on *ranks* (positions) but not on *items*. That is, the procedure randomly selects a subset of ranks and removes the items on these ranks, independently of the items themselves. In other words, an incomplete observation τ is obtained by projecting a complete ranking π on a random subset of positions $A \in 2^{[K]}$, i.e., the family (5) of distributions $p_\lambda(\cdot \mid \pi)$ is specified by a single measure on $2^{[K]}$. Or, stated more formally,

$$p_\lambda\big(\pi^{-1}(A) \mid \pi^{-1}\big) = p_\lambda\big(\sigma^{-1}(A) \mid \sigma^{-1}\big)$$

for all $\pi, \sigma \in \mathbb{S}^K$ and $A \subset [K]$, where $\pi^{-1}(A)$ denotes the projection of the ordering π^{-1} to the positions in A.

In the following, we make an even stronger assumption and assume observations in the form of (rank-dependent) pairwise comparisons. In this case, the coarsening is specified by probabilities

$$\left\{ \lambda_{i,j} \mid 1 \leq i < j \leq K,\ \lambda_{i,j} \geq 0,\ \sum_{1 \leq i < j \leq K} \lambda_{i,j} = 1 \right\},$$

where $\lambda_{i,j}$ denotes the probability that the ranks i and j are selected.

5.2.2   Likelihoods

Under the assumption of the PL model and rank-dependent pairwise comparisons as observations, the marginal likelihood for an observed set of pairwise comparisons $a_{i_n} \succ a_{j_n}$, $n \in [N]$, is given by

$$L_V(\theta, \lambda) = \prod_{n=1}^{N} \sum_{\pi \in \mathbb{S}_K, \pi(i_n) < \pi(j_n)} \lambda_{\pi(i_n), \pi(j_n)}\, \mathrm{pl}_\theta(\pi). \tag{9}$$

The corresponding expression for the face-value likelihood is

$$L_F(\theta) = \prod_{n=1}^{N} \frac{\theta_{i_n}}{\theta_{i_n} + \theta_{j_n}} = \prod_{i \neq j} \left(\frac{\theta_i}{\theta_i + \theta_j} \right)^{n_{i,j}}, \tag{10}$$

where $n_{i,j}$ denotes the number of times $a_i \succ a_j$ has been observed. This is the Bradely-Terry-Luce (BTL) model, which has been studied quite extensively in the literature [13].

Obviously, since the face-value likelihood is ignorant of the coarsening, we cannot expect the maximizer $\hat{\theta}$ of (10) to coincide with the true parameter θ. Interestingly, however, in our experimental studies so far, these parameters have always been asymptotically *comonotonic*, which is enough to recover the most probable ranking (2). That is, the face-value likelihood seems to yield reasonably strong estimates

$$\hat{\pi} = \arg_{k \in [K]} \text{sort} \left\{ \hat{\theta}_1, \dots, \hat{\theta}_K \right\} , \tag{11}$$

for sufficiently large samples, although the parameter $\hat{\theta}$ itself might be biased. The question whether or not this comonotonicity holds in general is still open. Yet, we could prove an affirmative answer at least under an additional assumption.

Theorem 1 *Suppose complete rankings to be generated by the PL model with parameters $\theta_1 > \theta_2 > \cdots > \theta_K$. Moreover, let the coarsening procedure be given by a rank-dependent selection of pairwise comparisons between items where λ satisfies the following condition:*

$$\lambda_{i,j} \geq \lambda_{i',j'}, \ if \ 1 \leq i \leq i' < j' \leq j \leq K.$$

Then, for an arbitrarily small $\epsilon > 0$ there exists $N_\epsilon \in \mathbb{N}$ such that, for every $N \geq N_\epsilon$ the maximizer $\hat{\theta}$ of (10) satisfies

$$\hat{\theta}_1 > \hat{\theta}_2 > \cdots > \hat{\theta}_K$$

with probability at least $1 - \epsilon$.

The proof of Theorem 1 can be derived from Lemmas 1 to 6.

Lemma 1 *Suppose complete rankings to be generated by the PL model with parameters $\theta_{i_1} > \theta_{i_2} > \cdots > \theta_{i_K}$. Given $i \neq j$, let $p_{i_k,i_l} = P(X \in E(a_{i_k} \succ a_{i_l})) = \frac{\theta_{i_k}}{\theta_{i_k} + \theta_{i_l}}$ denote the probability that a_{i_k} is preferred to a_{i_l}. Then, the following inequalities hold:*

$$p_{i_k,i_l} \geq p_{i_{k'},i_{l'}}, \ \forall k \leq k', l \geq l'.$$

Proof Straightforward, if we take into account that the mapping $f_c(x) = \frac{x}{x+c}$ is increasing on $\mathbb{R}_+$ for all $c > 0$ while $g_c : (x) = \frac{c}{c+x}$ is decreasing on $\mathbb{R}_+$.

Definition 1 Consider a complete ranking $\pi \in \mathbb{S}_K$, and let us consider two indices $i \neq j$. We define the (i,j)-*swap ranking*, $\pi_{i,j} : [K] \to [K]$, as follows: $\pi_{i,j}(k) = \pi(k)$, $\forall k \in [K] \setminus \{i,j\}$, $\pi_{i,j}(i) = \pi(j)$ and $\pi_{i,j}(j) = \pi(i)$.

Lemma 2 *Suppose complete rankings to be generated by the PL model* pl_θ. *Take* $i, j \in [K]$ *such that* $\theta_i > \theta_j$ *and* $\pi \in \mathbb{S}_K$. *Then:*

$$\pi(i) < \pi(j) \text{ if and only if } \mathrm{pl}_\theta(\pi) > \mathrm{pl}_\theta(\pi_{i,j}).$$

Proof Let us take an arbitrary ranking $\pi \in \mathbb{S}_k$ satisfying the restriction $\pi(i) < \pi(j)$ We can write:

$$\mathrm{pl}_\theta(\pi) = C_{i,j} \cdot \frac{\theta_{\pi^{-1}(\pi(i))}}{\sum_{s=\pi(i)}^{\pi(K)} \theta_{\pi^{-1}(s)}} \cdot \frac{\theta_{\pi^{-1}(\pi(j))}}{\sum_{s=\pi(j)}^{\pi(K)} \theta_{\pi^{-1}(s)}}$$

$$\mathrm{pl}_\theta(\pi_{i,j}) = C_{i,j} \cdot \frac{\theta_{\pi_{i,j}^{-1}(\pi_{i,j}(i))}}{\sum_{s=\pi_{i,j}(i)}^{\pi_{i,j}(K)} \theta_{\pi_{i,j}^{-1}(s)}} \cdot \frac{\theta_{\pi_{i,j}^{-1}(\pi_{i,j}(j))}}{\sum_{s=\pi_{i,j}(j)}^{\pi_{i,j}(K)} \theta_{\pi_{i,j}^{-1}(s)}},$$

where

$$C_{i,j} = \prod_{r \notin \{\pi(i), \pi(j)\}} \frac{\theta_{\pi^{-1}(r)}}{\theta_{\pi^{-1}(r)} + \theta_{\pi^{-1}(r+1)} + \cdots + \theta_{\pi^{-1}(K)}}$$

$$= \prod_{r \notin \{\pi_{i,j}(i), \pi_{i,j}(j)\}} \frac{\theta_{\pi_{i,j}^{-1}(r)}}{\theta_{\pi_{i,j}^{-1}(r)} + \theta_{\pi_{i,j}^{-1}(r+1)} + \cdots + \theta_{\pi_{i,j}^{-1}(K)}}.$$

According to the relation between π and $\pi_{i,j}$, we can easily check the following equality:

$$\sum_{s=\pi(i)}^{\pi(K)} \theta_{\pi^{-1}(s)} = \sum_{s=\pi_{i,j}(j)}^{\pi_{i,j}(K)} \theta_{\pi_{i,j}^{-1}(s)}$$

(In fact, both θ_i and θ_j appear in both sums). Therefore, $\mathrm{pl}_\theta(\pi) > \mathrm{pl}_\theta(\pi_{i,j})$ if and only if $\sum_{s=\pi(j)}^{\pi(K)} \theta_{\pi^{-1}(s)} < \sum_{s=\pi_{i,j}(i)}^{\pi_{i,j}(K)} \theta_{\pi_{i,j}^{-1}(s)}$. Furthermore, we observe that:

$$\sum_{s=\pi(j)}^{\pi(K)} \theta_{\pi^{-1}(s)} - \sum_{s=\pi_{i,j}(i)}^{\pi_{i,j}(K)} \theta_{\pi_{i,j}^{-1}(s)} = \theta_j - \theta_i,$$

and therefore $\mathrm{pl}_\theta(\pi) > \mathrm{pl}_\theta(\pi_{i,j})$ if and only if $\theta_j < \theta_i$.

Lemma 3 *If* $a > a'$ *and* $b > b'$ *then* $ab + a'b' > ab' + a'b$.

Proof Straightforward.

Lemma 4 *Suppose that* $\lambda_{k,l} \geq \lambda_{k',l'}$ *for all* k, l, k', l' *such that* $k \leq k', l \geq l', k < lk' <$ *l'. Suppose complete rankings to be generated according to a distribution p satisfying* $p(\pi) > p(\pi_{i,j})$ *for every* $\pi \in E(a_i \succ a_j)$, *for every pair* $i < j$.

Let $q_{i,j} = \sum_{\pi \in E(a_i > a_j)} p(\pi) \lambda_{\pi(i),\pi(j)}$, for all $i \neq j$ denote the probability of observing that a_i is preferred to a_j. Then

$$q_{i,j} > q_{i',j'} \text{ if } i \leq i' \text{ and } j \geq j'.$$

Proof We will divide the proof into three cases.

- Let us first prove that $q_{i,j} > q_{j,i}$, for all $i < j$. By definition, we have:

$$q_{i,j} = \sum_{\pi \in E(a_i > a_j)} p(\pi) \lambda_{\pi(i),\pi(j)} \text{ and } q_{j,i} = \sum_{\pi \in E(a_j > a_i)} p(\pi) \lambda_{\pi(j),\pi(i)}.$$

Furthermore, $E(a_j > a_i) = \{\pi_{i,j} : \pi \in E(a_i > a_j)\}$ and thus we can alternatively write:

$$q_{j,i} = \sum_{\pi \in E(a_i > a_j)} p(\pi_{i,j}) \lambda_{\pi_{i,j}(j),\pi_{i,j}(i)} = \sum_{\pi \in E(a_i > a_j)} p(\pi_{i,j}) \lambda_{\pi(i),\pi(j)}.$$

We easily deduce that $q_{i,j} > q_{j,i}$, $\forall i < j$ from the above hypotheses.

- Let us now prove that $q_{i,j} > q_{i+1,j}$, for all (i,j) with $i + 1 < j$. By definition we have:

$$q_{i,j} = \sum_{\pi \subset E(a_i > a_j)} p(\pi) \lambda_{\pi(i),\pi(j)} = \sum_{\pi \in \mathbb{S}_k} p(\pi) \alpha_{\pi(i),\pi(j)},$$

where $\alpha_{k,l} = \lambda_{k,l}$ for $k < l$ and $\alpha_{k,l} = 0$ otherwise. We can alternatively write:

$$q_{i,j} = \sum_{\pi \in E(a_i > a_{i+1})} p(\pi) \alpha_{\pi(i),\pi(j)} + \sum_{\pi \in E(a_{i+1} > a_i)} p(\pi) \alpha_{\pi(i),\pi(j)},$$

or, equivalently:

$$q_{i,j} = \sum_{\pi \in E(a_i > a_{i+1})} [p(\pi) \alpha_{\pi(i),\pi(j)} + p(\pi_{i,i+1}) \alpha_{\pi(i+1),\pi(j)}].$$

Analogously, we can write:

$$q_{i+1,j} = \sum_{\pi \in E(a_i > a_{i+1})} [p(\pi) \alpha_{\pi(i+1),\pi(j)} + p(\pi_{i,i+1}) \alpha_{\pi(i),\pi(j)}].$$

Now, according to the hypotheses, for every $\pi \in E(a_i > a_{i+1})$, $p(\pi) > p(\pi_{i,i+1})$ and $\alpha_{\pi(i),\pi(j)} > \alpha_{\pi(i+1),\pi(j)}$. Then, we can easily deduce from Lemma 3 that $q_{i,j} > q_{i+1,j}$.

- It remains to prove that $q_{i,j} > q_{i,j-1}$ for all (i,j) with $i + 1 < j$. The proof is analogous to the previous one.

Lemma 5 *For every $i \neq j$ let $n_{i,j}$ the number of times $a_i > a_j$ is observed in a sample of size N. Given $\epsilon > 0$ there exists $N_\epsilon \in \mathbb{N}$ such that for every $N \geq N_\epsilon$, the following equalities hold with probability at least $1 - \epsilon$:*

$$n_{i,j} \geq n_{i',j'}, \ \forall \, 1 \leq i \leq i' < j' \leq j \leq K.$$

Proof This result is a direct consequence of the WLLN and Lemma 4.

Lemma 6 *Consider the mapping* $g : \mathcal{M}_K(\mathbb{R}_+) \times \mathcal{M}_K(\mathbb{R}_+) \to \mathbb{R}_+$ *defined over the collections of pairs of K-square matrices of positive numbers as follows:*

$$g(\mathbf{r}, \mathbf{s}) = g((r_{i,j})_{i,j\in[K]}, (s_{i,j})_{i,j\in[K]}) = \prod_{i \neq j} r_{i,j}^{s_{i,j}}.$$

Suppose that the matrix $\mathbf{s} = (s_{i,j})_{i,j\in[K]}$ *satisfies the following restriction:*

$$s_{i,j} \geq s_{i',j'}, \ if \ i \leq i', j \geq j'.$$

Suppose that there exists $i^* \neq j^*$ *such that* $r_{i,j} \leq r_{i',i'}$ *for all* $(i,j), (i',j')$ *with:*

$$i \leq i', j \geq j', \ \{i,j\} \cap \{i^*,j^*\} \neq \emptyset \ and \ \{i',j'\} \cap \{i^*,j^*\} \neq \emptyset.$$

Consider the matrix $\mathbf{r}' = (r'_{i,j})_{i\in K, j\in[K]}$ *where:* $r'_{i,j} = r_{\sigma(i),\sigma(j)}$, *where* $\sigma \in \mathbb{S}_K$ *swaps* i^* *and* j^*, *i.e.,* $\sigma(i^*) = j^*$, $\sigma(j^*) = i^*$, $\sigma(k) = k$, $\forall k \in [K] \setminus \{i^*,j^*\}$.
Then $g(\mathbf{r}', \mathbf{s}) \geq g(\mathbf{r}, \mathbf{s})$.

Proof It is easy to prove that, under the above conditions, the ratio $\frac{g(\mathbf{r}',\mathbf{s})}{g(\mathbf{r},\mathbf{s})}$ is greater than 1.

5.2.3 Experiments

In order to compare the two approaches experimentally, synthetic data was produced by fixing parameters θ and λ and drawing N samples at random according to (4). Then, estimations $\hat{\theta}$ and $\hat{\pi}$ were obtained for both likelihoods, i.e., by maximizing (9) and (10). As a baseline, we also included estimates of θ assuming the coarsening λ to be known; to this end, (9) is maximized as a function of θ only. The three approaches are called MLM, FLM, and TLM, respectively.

The quality of estimates is measured both for the parameters and the induced rankings (11), in terms of the Euclidean distance between θ and $\hat{\theta}$, and in terms of the Kendall distance (relative number of pairwise inversions between items) between π and $\hat{\pi}$. The expectations of the quality measures were approximated by averaging over 100 simulation runs.

Here, we present results for a series of experiments with parameters $K = 4$, $\theta = (0.4, 0.3, 0.2, 0.1)$, and different assumptions on the coarsening:

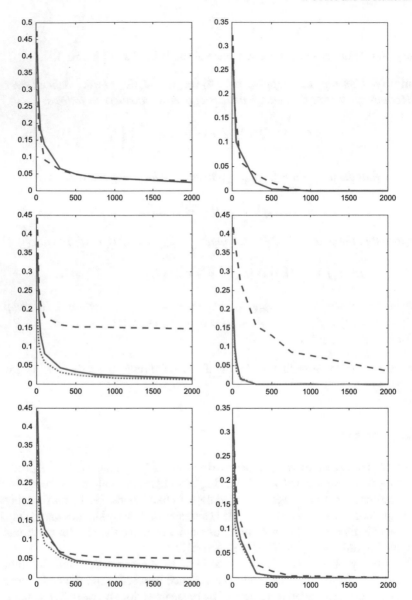

Fig. 1 Euclidean distance of parameter estimate $\hat{\theta}$ (left column) and Kendall distance of predicted ranking $\hat{\pi}$ (right column) for three experimental settings: uniform selection of pairwise comparisons (top), top-2 selection (middle), and rank-proportional selection (bottom). Curves are plotted in solid lines for MLM, dashed for FLM, and dotted for TLM

- In the first experiment, we set $\lambda_{1,2} = \cdots = \lambda_{3,4} = 1/6$. Thus, pairwise comparisons are selected uniformly at random. In this case, the face-value likelihood coincides with the likelihood of θ assuming the coarsening to be known, so this setting is clearly in favor of FLM (which, as already said, also coincides with TLM). Indeed, as can be seen in Fig. 1 (top), FLM yields very accurate estimates that improve with an increasing sample size. Nevertheless, MLM is not much worse and performs more or less on a par.
- In the second experiment, $\lambda_{1,2} = 1$ and $\lambda_{1,3} = \lambda_{1,4} = \cdots = \lambda_{3,4} = 0$. This corresponds to the top-2 setting, in which always the two items on the top of the ranking are observed. As expected, FLM now performs worse than MLM. As can be seen in Fig. 1 (middle, left), the parameter estimates of FLM are biased. Nevertheless, the estimation $\hat{\pi}$ is still decent (Fig. 1, middle, right) and continues to improve with increasing sample size.
- In the last experiment, items are selected with a probability inversely proportional their ranks: $\lambda_{i,j} \propto (8 - i - j)$. Thus, pairs on better ranks are selected with a higher probability than pairs on lower ranks. The results are shown in Fig. 1 (bottom). As can be seen, FLM is again biased and performs worse than MLM. However, the bias and the difference in performance are much smaller than in the top-2 scenario. This is hardly surprising, given that the coarsening λ in this experiment is less extreme than in the top-2 case. Instead, it is closer to the uniform coarsening of the first experiment, for which, as already said, FLM is the right likelihood.

6 Conclusion

This paper is meant as a first step toward learning from incomplete ranking data based on methods for learning from imprecise (set-valued) data. Needless to say, the scope of the paper is very limited, both in terms of the methods considered (inference based on the marginal and the face-value likelihood) and the setting analyzed (observation of pairwise comparisons based on the PL model with rank-dependent coarsening)— generalizations in both directions shall be considered in future work. Nevertheless, our results clearly reveal some important points:

- The arguably "correct" way of tackling the problem is complete inference about (θ, λ), i.e., about the complete data generating process, as done by MLM. While this approach will guarantee theoretically optimal results, it will not be practicable in general, unless the number of items is small or the parametrization of the coarsening process is simplified by very restrictive assumptions.
- Simplified estimation techniques such as FLM, which make incorrect assumptions about the coarsening or even ignore this process altogether, will generally lead to biased results.
- Yet, in the context of ranking data, one has to distinguish between the estimation of the parameter θ, i.e., the identification of the model, and the prediction of a related ranking π (typically the most probable ranking given θ, i.e., the mode of

the distribution). Indeed, the main interest often concerns π, while θ only serves an auxiliary purpose. As shown by the case of FLM, a biased estimation of θ does not exclude an accurate prediction of π, at least under certain assumptions on the coarsening process.

These observations suggest a natural direction for future work, namely the search for methods that achieve a reasonable compromise in the sense of being practicable and robust at the same time, where we consider a method robust if it guarantees a strong performance over a broad range of relevant coarsening procedures. Such methods should improve on techniques that ignore the coarsening, albeit at an acceptable increase in complexity.

References

1. Marden JI (1995) Analyzing and modeling rank data. Chapman and Hall, London, New York
2. Liu TY (2011) Learning to rank for information retrieval. Springer
3. Fürnkranz J, Hüllermeier E (2010) Preference learning. Springer
4. Ahmadi Fahandar M, Hüllermeier E, Couso I (2017) Statistical inference for incomplete ranking data: the case of rank-dependent coarsening. In: Proceedings ICML–2017, 34th international conference on machine learning, Sydney, Australia
5. Denoeux T (2011) Maximum likelihood estimation from fuzzy data using the EM algorithm. Fuzzy Sets Syst 183(1):72–91
6. Denoeux T (2013) Maximum likelihood estimation from uncertain data in the belief function framework. IEEE Trans Knowl Data Eng 25(1):119–130
7. Hüllermeier E (2014) Learning from imprecise and fuzzy observations: data disambiguation through generalized loss minimization. Int J Approx Reason 55(7):1519–1534
8. Plass J, Cattaneo M, Schollmeyer G, Augustin T (2016) Testing of coarsening mechanism: coarsening at random versus subgroup independence. In: Proceedings of SMPS 2016, 8th international conference on soft methods in probability and statistics. Springer, pp 415–422
9. Viertl R (2011) Statistical methods for fuzzy data. Wiley
10. Couso I, Dubois D. A general framework for maximizing likelihood under incomplete data (Submitted for publication)
11. Dawid AP, Dickey JM (1977) Likelihood and bayesian inference from selectively reported data. J Am Stat Assoc 72:845–850
12. Jaeger M (2005) Ignorability for categorical data. Ann Stat 33(4):1964–1981
13. Bradley RA, Terry ME (1952) The rank analysis of incomplete block designs I. The method of paired comparisons. Biometrika 39:324–345

Interval Type–2 Defuzzification Using Uncertainty Weights

Thomas A. Runkler, Simon Coupland, Robert John and Chao Chen

Abstract One of the most popular interval type–2 defuzzification methods is the Karnik–Mendel (KM) algorithm. Nie and Tan (NT) have proposed an approximation of the KM method that converts the interval type–2 membership functions to a single type–1 membership function by averaging the upper and lower memberships, and then applies a type–1 centroid defuzzification. In this paper we propose a modification of the NT algorithm which takes into account the uncertainty of the (interval type–2) memberships. We call this method the uncertainty weight (UW) method. Extensive numerical experiments motivated by typical fuzzy controller scenarios compare the KM, NT, and UW methods. The experiments show that (i) in many cases NT can be considered a good approximation of KM with much lower computational complexity, but not for highly unbalanced uncertainties, and (ii) UW yields more reasonable results than KM and NT if more certain decision alternatives should obtain a larger weight than more uncertain alternatives.

T.A. Runkler (✉)
Siemens AG, Corporate Technology, Otto-Hahn-Ring 6,
81739 Munich, Germany
e-mail: Thomas.Runkler@siemens.com

S. Coupland
Centre for Computational Intelligence, De Montfort University, The Gateway,
Leicester LE1 9BH, UK
e-mail: simonc@dmu.ac.uk

R. John · C. Chen
Laboratory for Uncertainty in Data and Decision Making (LUCID),
University of Nottingham, Wollaton Road,
Nottingham NG8 1BB, UK
e-mail: Robert.John@nottingham.ac.uk

C. Chen
e-mail: Chao.Chen@nottingham.ac.uk

© Springer International Publishing AG 2018
S. Mostaghim et al. (eds.), *Frontiers in Computational Intelligence*,
Studies in Computational Intelligence 739,
https://doi.org/10.1007/978-3-319-67789-7_4

47

1 Introduction

A type–1 fuzzy set A [1] is characterized by a membership function $u_A : X \to [0, 1]$ which quantifies the degree of membership of each element of X in A. Here we will always consider fuzzy sets over one–dimensional continuous intervals $X = [x_{\min}, x_{\max}]$. Type–1 defuzzification is a function d that maps a type–1 fuzzy set to one representative crisp value in X.

$$d(u(x)) \in X \tag{1}$$

Numerous methods for type–1 defuzzification have been proposed in the literature. For an overview see [2–4]. A set of desirable properties of type–1 defuzzification operators has been proposed in [5]. A popular method for type–1 defuzzification is the centroid function, which will be described in more detail in Sect. 2.

An interval type–2 fuzzy set [6–8] $\tilde{A}$ is characterized by two membership functions: a lower membership function $\underline{u}_{\tilde{A}} : X \to [0, 1]$ and an upper membership function $\overline{u}_{\tilde{A}} : X \to [0, 1]$, where

$$\underline{u}_{\tilde{A}}(x) \le \overline{u}_{\tilde{A}}(x) \tag{2}$$

for all $x \in X$. Interval type–2 fuzzy sets are known to be equivalent to interval–fuzzy sets [9, 10]. It was recently shown that type–2 fuzzy sets can be used to model risk in decision processes [11].

This paper deals with interval type–2 defuzzification, which is a function $\tilde{d}$ that maps an interval type–2 fuzzy set to one representative crisp value in X.

$$\tilde{d}(\underline{u}(x), \overline{u}(x)) \in X \tag{3}$$

A set of desirable properties of interval type–2 defuzzification operators has been proposed in [12]. A popular method for interval type–2 defuzzification is the Karnik–Mendel (KM) method [13], which will be described in more detail in Sect. 2.

Nie and Tan (NT) [14] have proposed an approximation of the KM method that first converts the interval type–2 membership functions to a single type–1 membership function by averaging the upper and lower memberships, and then applies the standard type–1 centroid defuzzification. We will describe this method in more detail in Sect. 3.

In this paper we propose a modification of the NT algorithm that takes into account the uncertainty of the (interval type–2) memberships. We call this method the *uncertainty weight* (UW) method. We compare the behavior of the KM, NT, and UW methods in extensive experiments motivated by fuzzy controller scenarios with different patterns of uncertainty.

This article is structured as follows: Sects. 2 and 3 briefly review the KM and NT interval type–2 defuzzification methods. Section 4 introduces the UW interval type–2 defuzzification method. Section 5 presents our experiments to evaluate and compare the KM, NT, and UW methods. Section 6 summarizes the conclusions of this work and points out some future research questions.

2 Karnik–Mendel Interval Type–2 Defuzzification

One of the most popular methods for type–1 defuzzification [2–4] is the centroid.

$$d_C(u(x)) = \frac{\displaystyle\int_{x_{\min}}^{x_{\max}} u(x) \cdot x \, dx}{\displaystyle\int_{x_{\min}}^{x_{\max}} u(x) \, dx} \tag{4}$$

The KM defuzzification [13] is an extension of the centroid defuzzification to interval type–2 fuzzy sets. For any given interval type–2 fuzzy set with the lower and upper membership functions $\underline{u}(x)$ and $\overline{u}(x)$, each embedded type–1 fuzzy set with the membership function $u(x)$ with

$$\underline{u}(x) \leq u(x) \leq \overline{u}(x) \tag{5}$$

will yield a centroid according to (4). The smallest and largest possible centroids of such embedded type–1 fuzzy sets are

$$\tilde{c}_l = \inf_{u(x) \in [\underline{u}(x), \overline{u}(x)]} \frac{\displaystyle\int_{x_{\min}}^{x_{\max}} u(x) \cdot x \, dx}{\displaystyle\int_{x_{\min}}^{x_{\max}} u(x) \, dx} \tag{6}$$

$$\tilde{c}_r = \sup_{u(x) \in [\underline{u}(x), \overline{u}(x)]} \frac{\displaystyle\int_{x_{\min}}^{x_{\max}} u(x) \cdot x \, dx}{\displaystyle\int_{x_{\min}}^{x_{\max}} u(x) \, dx} \tag{7}$$

These equations can be equivalently written as

$$\tilde{c}_l = \inf_{L \in [x_{\min}, x_{\max}]} \frac{\int_{x_{\min}}^{L} \overline{u}(x) \cdot x \, dx + \int_{L}^{x_{\max}} \underline{u}(x) \cdot x \, dx}{\int_{x_{\min}}^{L} \overline{u}(x) \, dx + \int_{L}^{x_{\max}} \underline{u}(x) \, dx} \tag{8}$$

$$\tilde{c}_r = \sup_{R \in [x_{\min}, x_{\max}]} \frac{\int_{x_{\min}}^{R} \underline{u}(x) \cdot x \, dx + \int_{R}^{x_{\max}} \overline{u}(x) \cdot x \, dx}{\int_{x_{\min}}^{R} \underline{u}(x) \, dx + \int_{R}^{x_{\max}} \overline{u}(x) \, dx} \tag{9}$$

The optimal switch points $L, R \in [x_{\min}, x_{\max}]$ can be found by the KM algorithm [13]. The result of the KM defuzzification is defined as the average of the smallest and largest possible centroids:

$$\tilde{d}(\underline{u}(x), \overline{u}(x)) = \frac{\tilde{c}_l + \tilde{c}_r}{2} \tag{10}$$

The next section provides the details of the Nie–Tan approach to interval type–2 defuzzification.

3   Nie–Tan Interval Type–2 Defuzzification

Nie and Tan (NT) [14] proposed an approximation of the KM method. The NT method first maps a given interval type–2 membership function to a type–1 membership function by averaging the upper and lower interval type–2 memberships.

$$u(x) = \frac{1}{2}(\underline{u}(x) + \overline{u}(x)) \tag{11}$$

Then NT computes the conventional type–1 centroid (4) of this type–1 membership function. Type–1 conversion (11) and computation of the type–1 centroid using (4) is computationally much cheaper than iteratively minimizing $\tilde{c}_l$ (8) and maximizing $\tilde{c}_r$ (9). Therefore, the NT method is a popular low effort approximation of the KM method.

Though NT is quite simple and straightforward, it is found that it may lose the information of uncertainty. Consider two data points x_1 and x_2 with the interval type–2 memberships $\underline{u}(x_1) = 0$, $\overline{u}(x_1) = 1$, $\underline{u}(x_2) = 0.5$, $\overline{u}(x_2) = 0.5$. For both data points the averaging function (11) will yield the same interval type–2 memberships $u(x_1) = u(x_2) = 0.5$, so both data points will have the same impact on the defuzzification result, although the membership of x_1 has a very high uncertainty reflected by the range of memberships from $\underline{u}(x_1) = 0$ to $\overline{u}(x_1) = 1$, and the membership of x_2 has a very low uncertainty reflected by the fact that the upper and lower memberships are equal, $\underline{u}(x_2) = \overline{u}(x_2) = 0.5$, so in this example the information about the uncertainty is lost by averaging the upper and lower memberships.

4   The Uncertainty Weight Method

In the NT method information about the uncertainty of the type–2 memberships is lost. To avoid this information loss we propose an uncertainty weight method. We define the degree of certainty of the memberships $\underline{u}(x)$ and $\overline{u}(x)$ as

$$w(x) = (1 + \underline{u}(x) - \overline{u}(x))^{\alpha} \tag{12}$$

with a suitable parameter $\alpha > 0$ that can be interpreted as a weight factor for the certainty. Smaller values of α will lead to a higher weight for medium uncertainties, and larger values of α will lead to a lower weight for medium uncertainties. In this paper we will always use $\alpha = 1$, which corresponds to a linear weight of the uncertainties. For our example above, Eq. (12) yields the certainty values $w(x_1) = 1 + 0 - 1 = 0$ and $w(x_2) = 1 - 0.5 + 0.5 = 1$, so data point x_1 is considered very uncertain, and data point x_2 is considered very certain. We want to reflect the (un)certainty of the different data points in defuzzification by using the certainty as a weight for each data point. This means that a relatively certain alternative has a large weight and that a relatively uncertain alternative has a small weight. Including the weights (12) in the averaging function (11) yields the weighted averaging function

$$u(x) = \frac{1}{2}(\underline{u}(x) + \overline{u}(x)) \cdot (1 + \underline{u}(x) - \overline{u}(x))^{\alpha} \tag{13}$$

Notice that for the NT method (11) we have for all x

$$\underline{u}(x) \leq u(x) \leq \overline{u}(x) \tag{14}$$

i.e. the effective membership $u(x)$ is bounded by the lower and upper type–2 memberships $\underline{u}(x)$ and $\overline{u}(x)$. For our proposed method even in the most certain situation $\underline{u}(x) = \overline{u}(x)$ Eq. (13) yields $u(x) = \overline{u}(x)$, so the effective membership can never exceed the upper type–2 membership. Consider however the example $\underline{u}(x) = 0.5$, $\overline{u}(x) = 1$, and $\alpha = 1$. Here, Eq. (13) yields $u(x) = 3/8$, so the effective membership can be lower than the lower type–2 membership, if the uncertainty is high. Here, the $u(x)$ are not really meant as the membership values of a type–1 fuzzy set, but merely as a defuzzification tool, and therefore we consider it acceptable that the type–1 fuzzy set described by $u(x)$ is not always part of the family of type–1 fuzzy sets that are compatible with the interval type–2 fuzzy set that is considered. We combine weighted averaging (13) with type–1 centroid defuzzification (4) and call this the *uncertainty weight* (UW) method. Just as NT, UW is computationally much cheaper than KM.

The main motivation for UW, however, is not only the computational cost, but also the explicit consideration of uncertainties.

5 Experiments

In this section we illustrate and compare the behavior of the KM, NT, and UW interval type–2 defuzzification methods. The results of the KM method are shown as solid lines, the results of the NT method are shown as dotted lines, and the results of the UW method are shown as dashed lines.

The considered examples are motivated by a fuzzy controller [15] with (for simplicity) two rules, Gaussian membership functions, and sum–product inference, so the fuzzy controller output is a weighted sum of two Gaussian membership functions. An application example for this setup is a controller of an autonomous vehicle avoiding an obstacle, where one rule triggers the action turn left and the other rule triggers turn right [16]. This scenario has received much attention in defuzzification because it requires a decision between possibly contradicting action alternatives, which may be viewed as an inconsistency in the rules base or may be considered as a challenge for the defuzzification process. We do not go into a deeper discussion of this aspect here, but simply consider this as a typical fuzzy controller scenario to evaluate and test our approach against the standard KM and NT methods. To mimic this scenario we consider the unit range $X = [0, 1]$ and construct interval type–2 membership functions by adding pairs of weighted Gaussian functions.

$$\underline{u}(x) = \underline{y}_1 \cdot e^{\left(-\frac{x-\mu_1}{2\sigma_1^2}\right)} + \underline{y}_2 \cdot e^{\left(-\frac{x-\mu_2}{2\sigma_2^2}\right)} \tag{15}$$

$$\overline{u}(x) = \overline{y}_1 \cdot e^{\left(-\frac{x-\mu_1}{2\sigma_1^2}\right)} + \overline{y}_2 \cdot e^{\left(-\frac{x-\mu_2}{2\sigma_2^2}\right)} \tag{16}$$

In our first set of experiments we investigate the effect of varying uncertainty on the results of the considered defuzzification methods. We keep one Gaussian constant and perform different variations of the uncertainty of the second Gaussian: difference between upper and lower memberships, the lower memberships only, and the upper memberships only.

In our first experiment we consider variations of the difference between the upper and the lower memberships. To do so, we set $\mu_1 = 1/4$, $\sigma_1 = 1/8$, $\mu_2 = 3/4$, $\sigma_2 = 1/8$, $\underline{y}_1 = 0.5$, $\overline{y}_1 = 1$, $\underline{y}_2 = 0.75 - \Delta/2$, $\overline{y}_2 = 0.75 + \Delta/2$, where the parameter Δ is varied in $[0, 0.5]$. The top left graph in Fig. 1 shows an example of this membership function(s) for $\Delta = 0.3$. Here, the uncertainty of the right Gaussian is a little smaller that the uncertainty of the left Gaussian. The different defuzzification results are marked by vertical lines. In this case, KM (solid) and NT (dotted) yield almost the same results, and UW (dashed) yields a slightly higher defuzzification result which takes into account the fact that the certainty on the right is higher than the certainty on the left. The top right graph in Fig. 1 shows the defuzzification results d for KM, NT, and UW as the uncertainty of the right Gaussian Δ is changed from 0 to 0.5. For $\Delta = 0.5$ both Gaussians are equal, and so for reasons of symmetry all three methods yield $\tilde{d} = 0.5$. As pointed out above, NT (dotted) ignores the different levels of uncertainty and therefore *always* yields the output $\tilde{d} = 0.5$. KM (solid) is only very slightly different from NT (dotted), so here NT is a good approximation of KM with a much lower computational effort. Only UW (dashed) takes into account the (un)certainty and yields a much higher output (closer to the right Gaussian) when the uncertainty of the right Gaussian is lower (for smaller values of Δ).

In our second experiment we consider variations of the lower memberships only, and set $\mu_1 = 1/4$, $\sigma_1 = 1/8$, $\mu_2 = 3/4$, $\sigma_2 = 1/8$, $\underline{y}_1 = 0.5$, $\bar{y}_1 = 1$, $\underline{y}_2 = h$, $\bar{y}_2 = 1$, where the parameter h is varied in $[0, 1]$. The second row of Fig. 1 shows the results of this experiment. On the left we see two Gaussians again, both with a maximum upper membership of one. The left Gaussian has a maximum lower membership of 0.5, and the right Gaussian has a maximum lower membership of h, in this case $h = 0.3$. Here, KM (solid vertical line) and NT (dotted vertical line) yield very similar results, and UW (dashed vertical line) yields a slightly smaller result. The right diagram in row 2 shows the results of the three methods for $h \in [0, 1]$. For $h \approx 0.5$ all three methods produce (almost) the same result around $\tilde{d} \approx 0.5$, as expected for

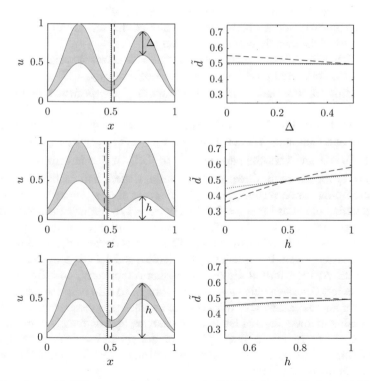

Fig. 1 Gaussians with different uncertainty patterns (solid: KM, dotted: NT, dashed: UW)

symmetry reasons. As the upper limit of the lower membership function decreases (lower h), UW (dashed) decreases the result most, then KM (solid), and NT (dotted) decreases the result least. For lower values of h, i.e. for quite unbalanced uncertainty patterns, KM (solid) and NT (dotted) yield significantly different results. This is an example, where NT does not approximate KM well. As the upper limit of the lower membership function increases (higher h), KM (solid) and NT (dotted) stay almost the same, but UW (dashed) increases the result much more, which reflects the higher certainty of the right Gaussian(s). For $h \in [0.5, 1]$ NT can again be considered a good low effort approximation of KM, but not for $h \in [0, 0.5]$, and UW better takes into account the varying uncertainty of the type–2 memberships.

In our third experiment we consider variations of the upper memberships only, and set $\mu_1 = 1/4$, $\sigma_1 = 1/8$, $\mu_2 = 3/4$, $\sigma_2 = 1/8$, $\underline{y}_1 = 0.5$, $\bar{y}_1 = 1$, $\underline{y}_2 = 0.5$, $\bar{y}_2 = h$, where the parameter h is varied in $[0.5, 1]$, see the third row of Fig. 1. On the left we see that the maximum upper membership of the right Gaussian is h, here $h = 0.7$. Also in this case KM (solid vertical line) and NT (dotted vertical line) yield almost the same results, but UW (dashed vertical line) yields a slightly higher result. The right diagram shows the results for $h \in [0.5, 1]$. For $h = 1$ we obtain the symmetric case again and all three methods yield $\tilde{d} = 0.5$. For $h < 1$ UW (dashed) stays almost constant at $\tilde{d} = 0.5$, because the reduction of the memberships is approximately compensated by the increased certainty. In contrast to that, KM (solid) and NT (dotted) are almost the same again and decrease with decreasing h. Again, NT is a good approximator for KM, but UW handles the varying uncertainties in an intuitively more reasonable way than KM and NT.

In our second set of experiments we investigate the effect of variations of horizontal widths σ_2, horizontal positions μ_2, and vertical scales $\bar{y}_2 = 2 \cdot \underline{y}_2$. In [5, 12] the corresponding transformations are called x–scaling, x–translation, and u–scaling, respectively.

The first row of Fig. 2 shows the effects of variations in the horizontal width σ_2 of the right Gaussian. We set $\mu_1 = 1/4$, $\sigma_1 = 1/8$, $\mu_2 = 3/4$, $\underline{y}_1 = 0.5$, $\bar{y}_1 = 1$, $\underline{y}_2 = 0.5$, $\bar{y}_2 = 1$ and vary the parameter σ_2 in $[0, 0.5]$. The left diagram shows the case $\sigma_2 = 1/16$, where all three methods (solid, dotted, and dashed vertical lines) yield almost the same results. The right diagram shows the results of the three defuzzification methods for $\sigma_2 = [0, 0.5]$. For $\sigma_2 = 1/8$ we have the symmetric case and all three methods yield $\tilde{d} = 0.5$. For smaller σ_2 all three methods yield almost the same results: As the width of the right Gaussian is decreased, the defuzzification result decreases as well. For smaller σ_2 KM (solid) and NT (dotted) yield almost the same results: As the width of the right Gaussian is increased, both Gaussians overlap and the left Gaussian gets larger memberships, so also here the defuzzification result becomes lower. For UW (dashed) the results stays close to $\tilde{d} \approx 0.5$ because the increasing memberships of the left Gaussian are approximately compensated by an increasing uncertainty (difference between upper and lower memberships of the left Gaussians). Also here, NT is a good approximator for KM, but UW better takes into account the uncertainties.

The second row of Fig. 2 shows the effects of variations in the horizontal position μ_2 of the right Gaussian. We set $\mu_1 = 1/4$, $\sigma_1 = 1/8$, $\sigma_2 = 1/8$, $y_{\underline{1}} = 0.5$, $\bar{y}_1 = 1$, $y_{\underline{2}} = 0.5$, $\bar{y}_2 = 1$ and vary the parameter μ_2 in $[0.5, 1]$. For $\mu_2 = 0.85$ (left diagram) all three methods (solid, dotted, and dashed vertical lines) yield almost the same results. The right diagram shows the results for $\mu_2 = [0.5, 1]$. For $\mu_2 = 3/4$ we have the symmetric case and all three methods yield $\tilde{d} = 0.5$. Also for all other values of $\mu_2 = [0.5, 1]$, all three methods yield almost the same results, so in this case both NT and UW are good approximators for KM.

The third row of Fig. 2 shows the effects of variations in the vertical scale of the right Gaussian. In contrast to the experiments in the third row of Fig. 1 we not only scale the upper membership function of the right Gaussian, $\bar{y}_2$, but also the lower membership function of the right Gaussian, $y_{\underline{2}}$, but keep the ratio between upper and lower memberships equal to 2, so that $\bar{y}_2 = 2 \cdot y_{\underline{2}}$. This simulates the situation that the first rule fires with strength 1 (yielding the left Gaussian) and the second rule fires with strength $h \in [0, 1]$ (yielding the right Gaussian), so in our experiments we can observe the behavior of the output when a rule fades out (or fades in). We set $\mu_1 = 1/4$, $\sigma_1 = 1/8$, $\mu_2 = 3/4$, $\sigma_2 = 1/8$, $y_{\underline{1}} = 0.5$, $\bar{y}_1 = 1$, $y_{\underline{2}} = h/2$, $\bar{y}_2 = h$

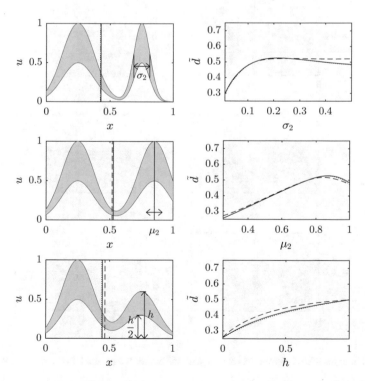

Fig. 2 Gaussians with different horizontal widths, horizontal positions, and vertical scales (solid: KM, dotted: NT, dashed: UW)

and vary the parameter h in $[0, 1]$. For $h = 0.6$ (left diagram) we have $\underline{y}_2 = 0.3$ and $\bar{y}_2 = 0.6$, and KM (solid vertical line) and NT (dotted vertical line) yield very similar results, whereas UW (dashed vertical line) yields a slightly higher result. The right diagram shows the results for $h \in [0, 1]$. For $h = 1$ we obtain the symmetric case and all three methods yield $\tilde{d} = 0.5$. For $h = 0$ the second Gaussian disappears and all three methods yield the center of the first Gaussian $\tilde{d} = 0.25$. The transition between the two extremes $h = 0$ (first rule completely active and second rule completely inactive) and $h = 1$ (both rules completely active) simulates a gradual increase of the firing strength of the second rule from zero to one. During this transition all three methods smoothly move from $\tilde{d} = 0.25$ at $h = 0$ to $\tilde{d} = 0.5$ at $h = 1$. KM (solid) and NT (dotted) yield almost the same results, but UW (dashed) yields slightly higher values, because the certainty of the right Gaussian is higher than the certainty of the left Gaussian. Here again, NT is a good approximator for KM but UW handles uncertainties in more plausible way.

We repeated the same experiments with triangular instead of Gaussian membership functions, where for comparability we chose the triangle widths as $4\sigma_1$ and $4\sigma_2$.

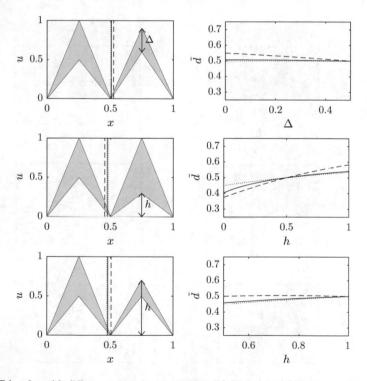

Fig. 3 Triangles with different uncertainty patterns (solid: KM, dotted: NT, dashed: UW)

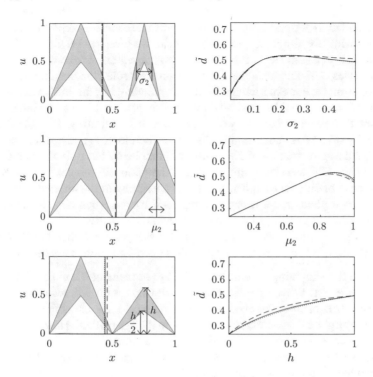

Fig. 4 Triangles with different horizontal widths, horizontal positions, and vertical scales (solid: KM, dotted: NT, dashed: UW)

$$\underline{u}(x) = \underline{y}_1 \cdot \max\left(0, 1 - \left|\frac{x - \mu_1}{2\sigma_1}\right|\right) + \underline{y}_2 \cdot \max\left(0, 1 - \left|\frac{x - \mu_2}{2\sigma_2}\right|\right) \quad (17)$$

$$\bar{u}(x) = \bar{y}_1 \cdot \max\left(0, 1 - \left|\frac{x - \mu_1}{2\sigma_1}\right|\right) + \bar{y}_2 \cdot \max\left(0, 1 - \left|\frac{x - \mu_2}{2\sigma_2}\right|\right) \quad (18)$$

Figure 3 shows the results of the triangle experiments corresponding to the Gaussian experiments shown in Fig. 1. The results of the triangular case are very similar to the results of the Gaussian case. Figure 4 shows the results of the triangle experiments corresponding to the Gaussian experiments shown in Fig. 2. Again, the results of the triangular case are very similar to the results of the Gaussian case.

6 Conclusions

We have proposed UW, a modification of the NT interval type–2 defuzzification method that takes into account the uncertainties of the (upper and lower) type–2 membership values. We performed extensive experiments comparing the standard

KM method with NT and UW. All experiments were motivated by fuzzy controller scenarios with (for simplicity) two rules, where we investigated the effect of different uncertainty patterns and of different horizontal widths, horizontal positions, and vertical scales of the membership functions on the defuzzification results.

To summarize, our experiments show the following: For the considered scenarios KM and NT mostly yield very similar results, except when parts of the interval type–2 membership function have very different levels of uncertainty. The computational complexity of KM is much higher than NT. Therefore, NT can often be considered a good approximation of KM with low complexity but only for well balanced uncertainties. UW also has a much lower computational complexity than KM, but in addition explicitly takes into account the uncertainty of the interval type–2 memberships, so it yields more reasonable results if more certain decision alternatives should obtain a larger weight than more uncertain alternatives.

This work is a first step in the explicit consideration of uncertainty in type–2 defuzzification. We have to leave many points open for future research, for example:

- We used the weighting scheme in Eq. (13) to implement the uncertainty weighting, with $\alpha = 1$. What are other equations or values of α will lead to a intuitively plausible treatment of the uncertainties in the type–2 memberships?
- We have applied the uncertainty weighting to the NT method. How could different levels of uncertainty be considered in the KM method?
- How does the behavior of all three methods change if we replace the centroid by other (type–1) defuzzification methods?

References

1. Zadeh LA (1965) Fuzzy sets. Inf Control 8:338–353
2. Roychowdhury S, Pedrycz W (2001) A survey of defuzzification strategies. Int J Intell Syst 16(6):679–695
3. Runkler TA (1997) Selection of appropriate defuzzification methods using application specific properties. IEEE Trans Fuzzy Syst 5(1):72–79
4. Saade JJ, Diab HB (2000) Defuzzification techniques for fuzzy controllers. IEEE Trans Syst Man Cybern Part B 30(1):223–229
5. Runkler TA, Glesner M (1993) A set of axioms for defuzzification strategies—towards a theory of rational defuzzification operators. In: IEEE international conference on fuzzy systems, San Francisco, pp 1161–1166
6. Liang Q, Mendel JM (2000) Interval type-2 fuzzy logic systems: theory and design. IEEE Trans Fuzzy Syst 8(5):535–550
7. Mendel JM, John RI, Liu F (2006) Interval type-2 fuzzy logic systems made simple. IEEE Trans Fuzzy Syst 14(6):808–821
8. Zadeh LA (1975) The concept of a linguistic variable and its application to approximate reasoning. Inf Sci 8:199–250, 301–357, 9:42–80
9. Gehrke M, Walker C, Walker E (1996) Some comments on interval valued fuzzy sets. Int J Intell Syst 11(10):751–759
10. Gorzałczany MB (1987) A method of inference in approximate reasoning based on interval-valued fuzzy sets. Fuzzy Sets Syst 21(1):1–17

11. Runkler TA, Coupland S, John R (2016) Interval type-2 fuzzy decision making. Int J Approx Reason 80:217–224
12. Runkler TA, Coupland S, John R (2015) Properties of interval type–2 defuzzification operators. In: IEEE international conference on fuzzy systems, Istanbul, Turkey
13. Karnik NN, Mendel JM (2001) Centroid of a type-2 fuzzy set. Inf Sci 132:195–220
14. Nie M, Tan WW (2008) Towards an efficient type–reduction method for interval type–2 fuzzy logic systems. In: IEEE international conference on fuzzy systems, Hong Kong, pp 1425–1432
15. Mamdani EH, Assilian S (1975) An experiment in linguistic synthesis with a fuzzy logic controller. Int J Man Mach Stud 7(1):1–13
16. Pfluger N, Yen J, Langari R (1992) A defuzzification strategy for a fuzzy logic controller employing prohibitive information in command formulation. In: IEEE international conference on fuzzy systems, San Diego, pp 717–723

Exploring Time-Resolved Data for Patterns and Validating Single Clusters

Frank Klawonn

Abstract Cluster analysis is often described as the task to partition a data set into subset—called clusters—so that similar data objects belong to the same cluster and data objects from different clusters are not very similar. However, partitioning the whole data set into clusters is often not the aim when clustering algorithms are applied. Instead, the main goal is sometimes to find a few "good" clusters containing a limited amount of data objects, while even the majority of data objects might not be assigned to any cluster, contradicting the principle of partitioning the data set into clusters. In this paper, we revisit a method called dynamic data assigning assessment clustering to discover and validate single clusters in a data set and extend the dynamic data assigning assessment approach to the context of time-resolved data.

1 What Cluster Analysis is Meant for and What it is used for

Cluster analysis is applied for different purposes. We extend the list given in [1] by an additional item.

- Finding clusters inherent in the data set under the assumptions that these cluster really exist. Most clustering algorithms are designed for this ideal purpose.
- Partitioning the data set into subset. Here it is not very important that the clusters are more or less well-separated. The partition of the data set is needed to reduce the complexity of the data set and handle the clusters separately. It is still important that the homogeneity criterion of cluster analysis is satisfied, i.e. that data objects in the same cluster are similar. But the heterogeneity criterion is dropped. Data objects from different clusters are allowed to be similar.

F. Klawonn (✉)
Biostatistics, Helmholtz Centre for Infection Research, Inhoffenstr. 7,
38124 Braunschweig, Germany
e-mail: f.klawonn@ostfalia.de

F. Klawonn
Department of Computer Science, Ostfalia University of Applied Sciences,
Salzdahlumer Str. 46/48, 38302 Wolfenbuettel, Germany

© Springer International Publishing AG 2018
S. Mostaghim et al. (eds.), *Frontiers in Computational Intelligence*,
Studies in Computational Intelligence 739,
https://doi.org/10.1007/978-3-319-67789-7_5

- Finding single clusters. Whereas standard clustering algorithms usually assume that the data set consists of clusters and a limited amount of noise data, here it is not important to cover a large fraction of the data by clusters. It is sufficient to find one or a few well-separated clusters and the majority of the data might not be assigned to any cluster.
- Data are already at least partly labelled by classes and one wants to verify whether the classes correspond roughly to clusters. Data from genomics and proteomics experiments [2] are a typical scenario for this clustering purpose. If data objects from the same class are scattered over all clusters, this might be an indication that something went wrong during the experiment and they data might be corrupted.
- Semi-supervised classification [3]. This situation is similar to the previous item, data objects are partly labelled by classes. The aim is now to assign the unlabelled data to classes. When the clusters corresponds well to the classes formed by the labelled data, it is reasonable to assign the an unlabelled data object to the class that has a (strong) majority in the corresponding cluster.
- Outlier detection. Here, the target is not the identification or separation of good clusters but the identification of single data objects that are isolated from most of the other data objects and are considered as outliers.

In this paper, we focus on time-resolved data as they are often generated in medical and biological studies that use high-throughput technologies like next generation sequencing, mass spectrometry or microarrays. Typically, so-called expression profiles over time of cellular molecular components like genes, proteins or metabolites are considered. The number of time points at which the expression values are taken, is usually very limit due to the costs and the lab and preparatory work required for such experiments. Sometimes, only three or four time points are measured in which case it is difficult to really talk about expression profiles over time. But with the advancement of technologies, experiments with ten or more time points can be seen. Time points need not be equidistant. The time points are usually chosen in such a way that in those periods where strong changes are expected, more measurements are taken whereas in periods where little is expected to change in the biological system, measurements are scarce. In contrast to the limited number of time points, the number of measured molecular components is usually very large, often greater than 1000. Usally, not all measured molecular components are involved in the observed biological process, so that many of the molecular components show very little expression over time and are then removed from further analysis, so that only molecular components are left over that show a significant expression at least at one of the time points. The aim is then to find groups (clusters) of molecular components that show similar expression profiles over time. The focus is put more on finding a few single clusters than assigning the expression profile of each molecular component to exactly one cluster [4, 5].

To tackle this problem, we propose an approach that allows to interactively explore the data set and identify and validate single clusters step by step.

We first briefly recall the idea of dynamic data assigning assessment clustering (DDAA) [6–8] and its extension [1] in the context of fuzzy clustering in Sect. 2.

Section 3 presents our ideas how DDAA can be applied to time-resolved expression data. In Sect. 4 we discuss a realistic example and also show the limitations before we conclude with a perspective on future work.

2 Short Revision of Dynamic Data Assigning Assessment Clustering in the Context of Fuzzy Clustering

We first recall basic notions from fuzzy clustering. The main reason is that fuzzy clustering provides interesting cluster validity measures that can be used in the context of DDAA. Here, it is sufficient to consider the simplest version, i.e. the fuzzy c-means algorithm (FCM) as it was introduced by Dunn and Bezdek [9, 10]. In Sect. 3 this algorithm will be modified to apply it to time-resolved data.

FCM clusters a data set $\{x_1, \ldots, x_n\} \subset \mathbb{R}^q$ by finding prototypes or cluster centres v_i for a given number of clusters c and assigning the data objects to clusters with membership degrees u_{ij}. FCM tries to minimise the objective function

$$f = \sum_{i=1}^{c} \sum_{j=1}^{n} u_{ij}^{w} d_{ij} \tag{1}$$

under the constraints

$$\sum_{i=1}^{c} u_{ij} = 1 \quad \text{for all } j = 1, \ldots, n \tag{2}$$

where $u_{ij} \in \{0, 1\}$ indicates whether data vector x_j is assigned to cluster i ($u_{ij} = 1$) or not ($u_{ij} = 0$). $d_{ij} = \| x_j - v_i \|^2$ is the squared Euclidean distance between data vector x_j and cluster prototype v_i. The minimisation of this objective function is carried out by an alternating optimisation schemes that alternatingly updates the prototypes v_i and the membership degree u_{ij}.

Davé [11] introduced an extension of FCM to cope with noise data. Noise clustering uses an additional noise cluster that does not have a prototype but a fixed large distance δ to all data objects. In this way, data objects that are far away from all clusters are assigned to the noise cluster with high membership degree.

DDAA [6–8] was originally introduced to identify clusters step by step and to remove noise data. DDAA is based on only one cluster and a noise cluster. It starts with a sufficiently large noise distance δ, so that all data objects are assigned to the single cluster with membership degree close to 1. Then the noise distance is decreased step by step, resulting in shifting more and more data from the single cluster to the noise cluster until all but one data object belong to the noise cluster. While data objects are shifted to the noise cluster, the prototype of the single cluster also changes its position. Tracking values of suitable indices like cluster validity during the reduction of the noise distance, DDAA provides interesting visualisation that can uncover the hidden cluster structure in the data. Original DDAA used the sum of the membership values of all data objects to the single cluster and change of this value to identify clusters and noise data. In [1], cluster validity measures were also used in the visualisation for DDAA. The advantage of fuzzy clustering here is that it provides good validity measures that are based on the membership degrees. An example for such a validity measures is the partition coefficient [10] (PC), defined by

$$\frac{\sum_{i=1}^{c} \sum_{j=1}^{n} u_{ij}^2}{n}.$$

The higher the value of the partition coefficient the better the clustering result. The highest value 1 is obtained, when the fuzzy partition is actually crisp, i.e. $u_{ij} \in \{0, 1\}$. The lowest value $1/c$ is reached, when all data are assigned to all clusters with the same membership degree $1/c$.

Another example is the partition entropy [10] (PE)

$$\frac{\sum_{i=1}^{c} \sum_{j=1}^{n} u_{ij} \ln(u_{ij})}{n}$$

inspired by the Shannon entropy. The smaller the value of the partition entropy, the better the clustering result. It should be noted that the first sum in PC and PE consists only of two summands: one for the single cluster and one for the noise cluster. Additional indices that are used for DDAA are the jumps, the ambiguity and the displacement that will be explained in the following section.

3   DDAA for Exploring Time-Resolved Data

In order to apply DDAA to time-resolved data, we first need a suitable distance measure. One could in principle use the (squared) Euclidean distance. But this would completely ignore the order of time points. If the time points were shuffled in all measured molecular components in the same way, the distance and therefore the result of the clustering would not change. Time points would be considered as attributes that have no connection. Here we propose to a modified distance measures that still yields a small distance between two expression profiles when they slightly shifted over time. In order to illustrate our concepts, we use a very small toy data sets shown in Fig. 1, consisting of seven different expression profiles.

Three expression profiles tend to increase and then to decrease again afterwards, forming one cluster. There is a second cluster of three other profiles that show an opposite behaviour. And there is one profile remaining almost constant at 0 that can be considered as noise. The three profiles in each cluster are slightly shifted in time. In order to take such shifts into account, we modify the (squared) Euclidean distance

Fig. 1 A toy example data
set for illustration purposes

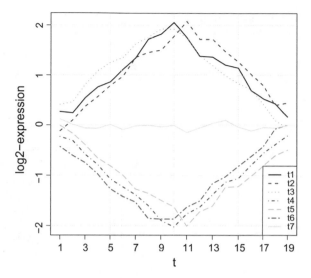

in FCM between a data object and a cluster prototype. We allow for a maximum
shift–say no more than two time points—to the left or to the right of the expression
profile and compute also the distances of the shifted expression profiles to the cluster
prototype. We also introduce a penalty for shifts, i.e. we multiply the distance based
on the shifted profile with a factor $a > 1$. More generally, the distance of an expression profile shifted by k time points is multiplied by the factor a^k. We then take the
smallest distance including penalty as the distance to the cluster. For updating the
prototype of the cluster, we need to know for which shift the distance of an expression profile including penalty was smallest. Instead of using the expression profile
directly, we use the shifted expression profile for updating the cluster prototype.

Figure 2 shows various indices how they evolve when the noise distance is
decreased. On the x-axis, the noise distance is indicated. On the y-axis, normalised
indices are indicated. These indices are always normalised in such a way that they
never exceed 1. The curves should be read from right to left because one starts with
a large noise distance that is decreased step by step.

The Sum curve shows the sum of the rounded membership degrees to the single
cluster. dSum is the change of the sum curve. PC and PE are the partition coefficient
and the partition entropy respectively. The Jumps curve shows the changes of the
position of the prototype of the single cluster, normalised by its highest value. The
Amb curve in Fig. 1—for ambiguity—shows the proportion of ambiguous membership degrees. Here we have defined a membership degree to be ambiguous if it is
between 0.3 and 0.7. The Displ curve—for displacement—shows the displacement
of the prototype from its initial starting position, normalised by its highest value.

In the very left part of Fig. 1 where the noise distance is close to 0, one can observe
interesting changes in most of the indices at a noise distance of $\delta \approx 0.25$. With this
noise distance, all but one of the seven expression profiles are moved to the noise

Fig. 2 DDAA visualisation
for the data set in Fig. 1

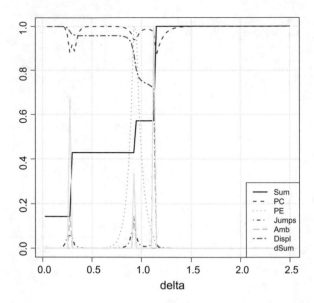

cluster. This is not a very interesting point. This will always happen because the last
remaining profile in the cluster will then be (almost) identical to the prototype of the
cluster and will therefore have distance of (almost) 0 to the prototype.

The more interesting point in Fig. 2 can be observed at a noise distance of $\delta \approx$
0.9. At this point, all expression profiles except one clusters are moved to the noise
cluster. This would be an interesting candidate to be a proper cluster in our data
set. Figure 3 where the course of the membership degrees [1] depending on the noise
distance are shown confirms this observation. For noise distances less than 0.9 the
three upper expression profiles in Fig. 1 are assigned to the actual DDAA cluster
with a membership degree greater than 0.5, whereas the other profiles are assigned
to the noise clusters. At a noise distance of roughly 0.9, the expression profiles still
belongs to the DDAA cluster and at a noise distance of roughly 1.1 and larger all
expression profiles form the DDAA cluster.

One would then remove this first cluster of expression profiles from the data set
and apply DDAA again to the remaining data set. One would then find the second
cluster in the lower part of Fig. 1 in a similar way.

---

[1]We have added at small random constant to each membership curve in order to better separate
the curves in the graphic. The actual membership degrees without the constant in graphic never
exceed 1.

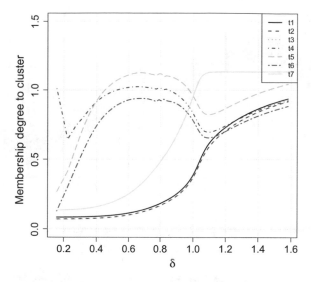

Fig. 3 Course of the (slightly modified) membership degrees for DDAA applied to the data set in Fig. 1

4 An Example

In this section, we discuss a more realistic artificial example data set that is similar in terms of the clusters as the data set in Fig. 1. It consists of two clusters similar to the ones in Fig. 1. There is one cluster with 100 expression profiles with noise that tend to go up and then go down again. The second cluster consists of 50 expression profiles with noise with the opposite behaviour. The two clusters are shown in Fig. 4. The majority of the data set consists of 200 noise expression profiles in the form of random walks normalised to an absolute peak of 2. The noise profiles are shown in Fig. 5 on the left hand side, all expression profiles together including the two clusters in Fig. 4 are plotted in Fig. 5 on the right hand side.

The result of the application of the DDAA algorithm is shown in Fig. 6 where the indices show an interesting behaviour at a noise distance of slightly less than 0.5. In very similar way as in the toy example in the previous section, this is the noise distance where the cluster on the left of Fig. 4 belongs to the DDAA cluster and most of the other expression profiles are assigned to the noise cluster. Taking at the membership degrees to the cluster at a noise distance of 0.5, we find 109 expression profiles with a membership degree larger than 0.5 consisting of the 100 expression profiles from the cluster shown in Fig. 4 (left). The other nine expression profiles come from the random walks that by chance have a similar profile as the other 100 expression profiles. The nine expression profiles originating from the random walks are shown in Fig. 7.

If one suspects or expects certain expression profiles, on can also search directly for them in the following way—a kind of reverse version of the DDAA algorithm.

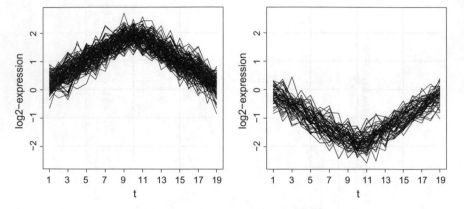

Fig. 4 Two clusters of expression profiles in a larger example data set

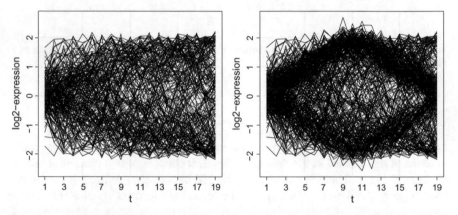

Fig. 5 Noise profiles (left) in the form of random walks and all expression profiles including the clusters shown in Fig. 4

Assume we take an expression profile that reflects the shape of the expression profiles in the cluster shown in Fig. 4 (right). Here the DDAA approach is used in a reverse manner. It is started with noise distance of almost 0 and the ideal prototype expression profile one is searching for. In this case, we would take an ideal expression profile that starts at 0 at the initial time point, decreases linearly to -2 until the middle of the time course and increases back to 0 linearly until the end of the time course. Increasing the noise distance step by step one obtains the profile of the DDAA indices as illustrated in Fig. 8. Note that in this case the diagram should be read from left to right because now the noise distance is increased step by step.

The interesting change seems to happen when the noise distance exceeds a value of over 0.3. Until this point, the DDAA cluster seems to be quite stable. And indeed, when we take a look at the membership degrees for a noise distance of $\delta = 0.3$, we obtain 49 profiles with a membership degree larger than 0.5 to the DDAA cluster. The

Fig. 6 DDAA visualisation for the data set shown in Fig. 5 (right)

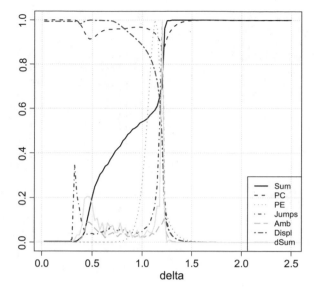

Fig. 7 Random walk expression profiles that are "wrongly" assigned to the expression profiles in Fig. 4 (left)

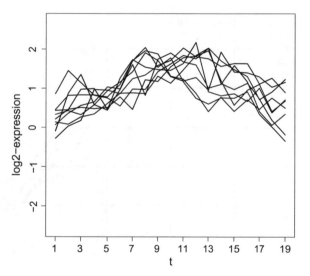

cluster on the right hand side of Fig. 4 contributes 46 profiles and three profiles come from the random walk profiles. Figure 9 the expression profiles that were missed from the cluster in Fig. 4 (right) (full black lines) and those that were included from the random walk profiles (red dotted lines).

Fig. 8 "Reverse" DDAA
starting with the ideal
expression profile for the
cluster in Fig. 4 (right)

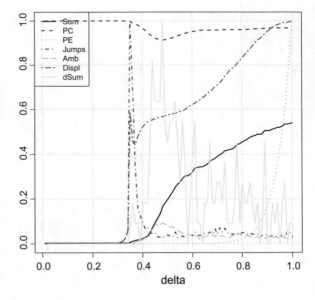

Fig. 9 Expression profiles
that are not included (full
black lines) or wrongly
included (red dotted lines,
originating from the random
walk expression profiles) in
the cluster with the ideal
expression profile for the
cluster in Fig. 4 (right)

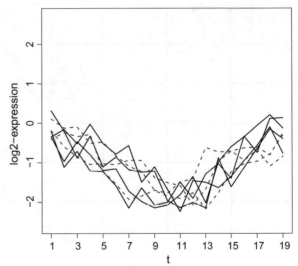

5 Conclusions

We have proposed an approach to discover single clusters of profiles in time resolved
data as they are often found in expression measurements over time in genomics,
proteomics or metabolomics. It is an exploratory method that can point to single
"good" clusters step by step. When one cluster is found it should be removed from
the data set and the same procedure should be applied again the reduced data set.

In the future, we will verify our approach with real data and also consider other distance measures, for instance based on Spearman's rank correlation coefficient ρ. If $1 - \rho$ is used as the distance for clustering, the prototype update equation needs to be modified. Since ρ only takes the order of the values into account, not their magnitude, it is sufficient to replace all values by their order, i.e. an expression profile like $(0.2, 0.8, 0.7, 0.9, 0.5)$ would be replaced by $(1, 5, 3, 2, 4)$ because the first value is the smallest value, the fifth value is the second smallest value and so on. Then our approach can be directly applied to this modified data set.

References

1. Klawonn F (2106) Exploring data sets for clusters and validating single clusters. Procedia Comput Sci 96 (1381–1390)
2. Kerr G, Ruskin H, Crane M (2008) Techniques for clustering gene expression data. Comput Biol Med 38(3):383–393
3. Bair E (2013) Semi-supervised clustering methods. Wiley Interdiscip Rev Comput Stat 5:349–361
4. Ernst J, Nau G, Bar-Joseph Z (2005) Clustering short time series gene expression data. Bioinform 21:159–168
5. Ernst J, Bar-Joseph Z (2006) Stem: a tool for the analysis of short time series gene expression data. BMC Bioinform 7:191. doi:10.1186/1471-2105-7-191
6. Georgieva O, Klawonn F (2006) Evolving clustering via the dynamic data assigning assessment algorithm. In: Proceedings IEEE international symposium on evolving fuzzy systems. pp 95–100
7. Georgieva O, Klawonn F (2006) Cluster analysis via the dynamic data assigning assessment algorithm. Inf Technol Control 2:14–21
8. Georgieva O, Klawonn F (2008) Dynamic data assigning assessment clustering of streaming data. App Soft Comput 8:1305–1313
9. Dunn J (1973) A fuzzy relative of the isodata process and its use in detecting compact well-separated clusters. Cybern Syst 3(3):32–57
10. Bezdek J (1981) Pattern recognition with fuzzy objective function algorithms. Plenum Press, New York
11. Davé RN (1991) Characterization and detection of noise in clustering. Pattern Recognit Lett 12:406–414

Interpreting Cluster Structure in Waveform Data with Visual Assessment and Dunn's Index

Sara Mahallati, James C. Bezdek, Dheeraj Kumar, Milos R. Popovic and Taufik A. Valiante

Abstract Dunn's index was introduced in 1974 as a way to define and identify a "best" crisp partition on n objects represented by either unlabeled feature vectors or dissimilarity matrix data. This article examines the intimate relationship that exists between Dunn's index, single linkage clustering, and a visual method called iVAT for estimating the number of clusters in the input data. The relationship of Dunn's index to iVAT and single linkage in the labeled data case affords a means to better understand the utility of these three companion methods when data are crisply clustered in the unlabeled case (the real case). Numerical examples using simulated waveform data drawn from the field of neuroscience illustrate the natural compatibility of Dunn's index with iVAT and single linkage. A second aim of this note is to study customizing the three methods by changing the distance measure from Euclidean distance to one that may be more appropriate for assessing the validity of crisp clusters of finite sets of waveform data. We present numerical examples that support our assertion that when used collectively, the three methods afford a useful approach to evaluation of crisp clusters in unlabeled waveform data.

S. Mahallati (✉) · M.R. Popovic · T.A. Valiante
Institute of Biomaterials and Biomedical Engineering,
University of Toronto, Toronto, Canada
e-mail: sara.mahallati@utoronto.ca

S. Mahallati · M.R. Popovic
Toronto Rehabilitation Institute, University Health Network,
Toronto, Canada

T.A. Valiante
Krembil Research Institute, University Health Network,
Toronto, Canada

D. Kumar
Lyles School of Civil Engineering,
Purdue University, West Lafayette, IN, USA

J.C. Bezdek
Computer Science and Information Systems Departments,
University of Melbourne, Melbourne, Australia

© Springer International Publishing AG 2018
S. Mostaghim et al. (eds.), *Frontiers in Computational Intelligence*,
Studies in Computational Intelligence 739,
https://doi.org/10.1007/978-3-319-67789-7_6

73

1 Introduction

Let $O = \{o_1, \ldots o_n\}$ denote any set of n objects (hockey teams, airplanes, epilepsy patients, etc.). Two kinds of numerical data are used to represent O. Numerical object data (feature vector data) has the form $X = \{\mathbf{x}_1, \ldots \mathbf{x}_n\} \subset \mathfrak{R}^p$, where the coordinates of x_i are feature values or attributes (e.g., weight, length, intensity, etc.) describing various properties of object o_i, $1 \leq i \leq n$. Regard the vectors as column vectors.

A second form of data is numerical relational data, which consists of pair-wise dissimilarities represented by an $n \times n$ matrix $D = [d_{ij}] = [diss(o_i, o_j) : 1 \leq i, j \leq n]$. Following the terminology used in [1], we regard cluster analysis in X or D as a set of three canonical problems: (P1) pre-clustering assessment of tendency (do the data have clusters? If yes, at what value of c in $\{2, 3, \ldots, n-1\}$?), (P2) partitioning the data (how to find the c clusters?), and (P3) post-clustering cluster validity (are the c clusters found useful?). This study concerns itself with problems (P1) and (P3). There is a very fine line between the distinctions made for these two problems.

The non-degenerate crisp c-partitions (i.e. no zero rows corresponding to empty clusters) of n objects are matrices in the following set:

$$M_{hcn} = \{U \subset \mathfrak{R}^{cn} : 1 \leq i \leq c, 1 \leq k \leq n : u_{ik} \in \{0, 1\};$$

$$\sum_{i=1}^{c} u_{ik} = 1 \forall k; \quad \sum_{k=1}^{n} u_{ik} > 0 \forall i\} \tag{1}$$

An equivalent way to represent the c clusters in a set X in terms of the c crisp subsets of it (the clusters) is

$$U \in M_{hcn} \leftrightarrow X = \bigcup_{i=1}^{c} X_i; X_i \cap X_j = \emptyset|_{i \neq j}$$

Cluster validity (validation problem (P3)) comprises computational models and algorithms that identify a "best" member amongst a set of *candidate partitions* $CP = \{U \in M_{hcn} : c_m \leq c \leq c_M\}$ of $O = \{o_1, o_2, \ldots o_n\}$. How do we identify a "best" partition U (and concomitant value of c) in CP? One approach is through the use of scalar measures of partition quality that attempt to select a best candidate. There are many such measures (enough equations to last a lifetime!). We call these measures *cluster validity indices* (CVIs).

There are two general approaches to validation: visual and non-visual. Perhaps the most important distinction is whether the partition U is crisp or soft. A second distinction is based on the type of data that is clustered: object data (X) and relational data (R). A third characteristic will be whether the validation method is internal (uses only the information available from the algorithmic outputs), or external (uses "outside" information such as a reference or ground truth partition). This article is confined to internal CVIs, but we will use crisp clusters corresponding to ground truth partitions of simulated waveform data as the basis of our numerical examples.

We will combine visual assessment of potential cluster structure in data (problem P1) with cluster validity in crisp (ground truth) clusters of labeled data to develop a facility for evaluation of potential structure in unlabeled waveform data. For this study we will use only feature vector data that represent finite time series (waveforms) for the numerical experiments.

The outline of the paper is as follows. Section 2 discusses related work and applications. *Dunn's index*(DI) is defined and discussed in Sect. 3. Section 4 describes and illustrates the iVAT algorithm. Section 5 reviews single linkage clustering. Section 6 focuses on the relationship between iVAT, single linkage, and Dunn's index. Section 7 introduces our computational protocols. In this section we define and analyze a set of simulated waveform generators which are used to build test cases of labeled data with $c = 2$ clusters of waveforms for study in subsequent sections. We also explore customizing Dunn's index for waveform data in Sect. 7. Section 8 revisits single linkage clustering and shows the relationship between Dunn's index and a measure of partition accuracy. Finally, Sect. 9 offers our conclusions, and some suggestions for further research. Appendices 1–3 give pseudocode for three algorithms used in the experiments.

2 Related Work

Cluster validity is a very large topic. Many books that cover cluster analysis contain at least one chapter on cluster validity [1–4]. Surveys on crisp cluster validity indices (CVIs) that compare various validation schemes in one way or another began to appear in the 1980s [5]. Milligan and Cooper [6] compared 30 internal cluster validity tests (which they called "stopping rules") using partitions generated by four hierarchical clustering methods, and their paper is considered the classic reference on best-c studies of internal CVIs. Gurrutxaga et al. [7] present a very thorough critique of the Milligan and Cooper's "best c" methodology.

Dimitriadou et al. [8] made a highly specialized, nicely written survey of 15 internal CVIs in 2002. Vinh et al. [9] presented a best-c study of similarity measures and distance based external CVIs that compare pairs of crisp partitions using information-theoretic measures. They identify a total of 26 measures that are subdivided into 10 similarity measures and 16 distance measures. Arbelaitz et al. [10] published a very ambitious comparison of 30 internal min-optimal or max-optimal CVIs for crisp c-partitions that channels the Milligan-Cooper style. Three crisp clustering algorithms were used to populate CP: It would be misleading to assert that the conclusions in this paper are right or wrong. Indeed, the terms "good CVIs" and "bad CVIs" are oxymorons in the field of cluster validity. But this is a strong, well thought out, comprehensive approach to the business of CVI comparisons that should be studied by anyone with serious intentions to work in the field of cluster validation.

3 Dunn's Index

Dunn's index is an internal CVI for assessing cluster validity for crisp c-partitions of a data set [11]. This index is based on the geometrical premise that "good" sets of clusters are compact (dense about their means) and well separated (from each other). To quantify this index Dunn let X_i and X_j be non empty subsets of $\mathfrak{R}^p$, and let $d : \mathfrak{R}^p \times \mathfrak{R}^p \mapsto \mathfrak{R}^+$ be any metric on $\mathfrak{R}^p \times \mathfrak{R}^p$. Dunn used the standard definitions of the diameter $\triangle$ of X_k and the set distance δ between X_i and X_j,

$$\Delta(X_k|d) = \underbrace{max}_{\mathbf{x},\mathbf{y} \in X_k} \{d(\mathbf{x},\mathbf{y})\}, \tag{2}$$

$$\delta(X_i, X_j|d) = \underbrace{min}_{\substack{\mathbf{x} \in X_i \\ \mathbf{y} \in X_j}} \{d(\mathbf{x},\mathbf{y})\} = \delta_{SL}(X_i, X_j|d). \tag{3}$$

The geometric meaning of these definitions is illustrated for the case that d is the Euclidean distance (d = ED) in Fig. 1. The subscript "SL" attached to δ in the right side of (3) signifies that this is the set distance used by the hierarchical single linkage (SL) clustering algorithm [1–4]. This is an important point for our work, because this distance binds Dunn's index to the iVAT visualization method discussed in Sect. 4 and to single linkage clustering discussed in Sect. 8.

For any partition $U \leftrightarrow X = X_1 \cup \dots X_i \cup \dots X_c$, Dunn defined the separation index of U as

$$\beta_D(U|d) = \frac{\underbrace{min}_{1 \le i \le c} \left\{ \underbrace{min}_{1 \le j \ne i \le c} \{\delta(X_i, X_j|d)\} \right\}}{\underbrace{max}_{1 \le k \le c} \{\Delta(X_k|d)\}} \tag{4}$$

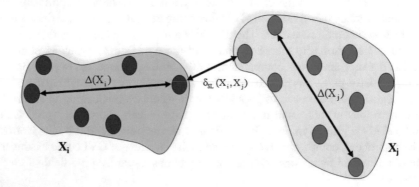

Fig. 1 Set distance (δ) and diameter (Δ) of X_i and X_j with respect to metric d

Dunn did not restrict the metric d used in (4), and hence, in (2) or (3) as well. This function was specified as an arbitrary metric on any real vector space $V(\mathfrak{R})$, having the usual three properties for $\mathbf{u}, \mathbf{v}, \mathbf{w} \in V(\mathfrak{R})$:

- M1: $d(\mathbf{u}, \mathbf{v}) > 0$ if $\mathbf{u} \neq \mathbf{v}$; otherwise $d(\mathbf{u}, \mathbf{v}) = 0$ (positive-definiteness),
- M2: $d(\mathbf{u}, \mathbf{v}) = d(\mathbf{v}, \mathbf{u})$ (symmetric),
- M3: $d(\mathbf{u}, \mathbf{w}) \leq d(\mathbf{u}, \mathbf{v}) + d(\mathbf{v}, \mathbf{w})$ (triangle inequality).

The notation $|d$ for the arguments of the functions in (2), (3) and (4) is read "given d". This emphasizes the fact that the theory and application of Dunn's index holds for any metric (d) on the input space.

Following many subsequent authors, we refer to β_D in (4) as Dunn's index (DI). The most common metric for d in the numerator and denominator of β_D (i.e., in (2) and (3) is the *Euclidean distance* (ED), but there are many other choices [12]. One objective of the present work is to study the behavior of Dunn's index when (d) is replaced by dissimilarity measures such as the *shape based distance* (SBD) for waveforms [13] or *dynamic time warping* (DTW) [14] which are specifically designed for waveform data. To distinguish them from the *generalized Dunn's indices* (gDIs) discussed in [12], we will call this new group of measures *customized Dunn's indices* (cDIs).

The diameter $\Delta(X_k|d)$ in (2) which appears in the denominator of (4) is a measure of scatter volume for cluster X_k. Compact clusters will have smaller diameters than ones that are more dispersed about their mean vector, so a set of clusters is relatively compact when the largest of its c diameters is small. So, for the denominator, the smaller the better. That is, when the biggest diameter in the X_i is small, the clusters in U are relatively compact.

The quantity $\delta(X_i, X_j|d)$ which appears in the numerator of (4) is the SL distance with respect to (d) at Eq. (3) between the crisp clusters X_i and X_j in U, and hence, the larger this distance, the more well separated are X_i and X_j. Taking the double minimum in the numerator identifies the pair of clusters that are least well separated. So, the c clusters becomes more well separated as the numerator grows.

Thus, the geometric objective of Dunn's index is to maximize intercluster distances (big numerators) whilst minimizing intracluster distances (small denominators). In other words, large values of β_D intuitively correspond to good clusters. Hence, the partition U^* that maximizes β_D over a set of candidate partitions in CP is taken as the optimal set of clusters. Consequently, β_D is called a "max-optimal" internal CVI. The range of β_D is $(0, \infty)$. Notice that β_D is not defined when $c = 1$ on the $1 \times n$ vector 1_n (the unique partition of X which denotes that all n points are in one cluster). And β_D is also undefined for the case $c = n$, when the n members of X are regarded as singleton clusters, whose unique partition is the $n \times n$ identity matrix I_n.

Dunn called a partition $U \in M_{hcn}$ *compact and separated* (CS) relative to d if and only if the following property is satisfied: for all s, q and r with $q \neq r$, any pair of points $\mathbf{x}, \mathbf{y} \in X_s$ are closer together (with respect to d) than any pair $(\mathbf{u}, \mathbf{v})$ with $\mathbf{u} \in X_q$ and $\mathbf{v} \in X_r$. Dunn proved that X can be clustered into a CS c-partition with respect to d if and only if there is a $U \in M_{hcn}$ for which his index is greater than 1.

Here is Dunn's Theorem:

$$X \subset \mathfrak{R}^p \text{ has a CS partition with respect to metric d} \Leftrightarrow \underbrace{max}_{U \in M_{hcn}} \{\beta_D(U|d)\} > 1 \quad (5)$$

This is a nice theoretical result, but in practice, it is quite difficult (impossible, really) to verify that a given input data set can be partitioned into CS clusters, because M_{hcn} is finite, but very very large. The exact cardinality of M_{hcn} is $|M_{hcn}| = \frac{1}{c!} \sum_{j=1}^{c} \binom{c}{j} (-1)^{c-j} j^n$. For $c \ll n$, the last term dominates this sum, which yields the approximation $|M_{hcn}| = \frac{c^n}{c!}$.

Consequently, computing Dunn's index over all of M_{hcn} is impractical for all but trivial values of c and n. The DI for *a given c-partition*, however, is easily computed with (4), and if it happens to be greater than 1, its clusters are said to be compact and separated in the sense of Dunn.

4   Visualization of High Dimensional Data

Returning to the data set $X = \{\mathbf{x}_1, \dots \mathbf{x}_n\} \subset \mathfrak{R}^p$, if $p > 3$ we cannot make a direct visual assessment of it to determine whether it contains what humans might perceive as cluster structure. In this case $\mathfrak{R}^p$ is called the *upspace*. The data can be mapped to a *downspace* $Y \subset \mathfrak{R}^q$ by a feature extraction function $\phi : \mathfrak{R}^p \mapsto \mathfrak{R}^q$ in any number of ways. There are two reasons to do this: (i) dimensionality reduction improves time complexity and efficiency for clustering and classification algorithms; (ii) when the method results in 1D, 2D or 3D vectors, we can display and see them. It is important to note that there is always a loss or distortion of structural information in such a transformation. Any mapping from the upspace to a downspace becomes a trade-off between computational benefits and performance goals. Visual analysis can give us clues about structure in the upspace data, but this can be very misleading: we cannot "see" the upspace data, and data structure can be very different in the upspace than it seems to in a downspace (like Plato's Allegory of the cave).

For example, Figs. 3, 4, 5, 6, 7, 8 and 9 in [15] show 2D representations of a labeled data set in an upspace with p = 9601 dimensions made by eight different extraction methods. The eight 2D scatter-plots of this data are so different from each other that you would never guess they all purport to represent the same upspace data. The takeaway lesson from this is that depending on any downspace projection for visualization of upspace data can lead to serious misinterpretations of cluster structure in it.

How can we reduce the possibility of misinterpreting possible cluster structure in upspace data when viewing it in a 2D or 3D downspace? There are a number of imaging techniques that can be applied directly to the upspace data before clustering it. Here we describe the VAT/iVAT method, which views the vectors in X as n vertices in a complete graph whose edge weights are the distances (or dissimilarities) between the vertices, computed by any metric on the upspace.

The *visual assessment of tendency* (VAT) algorithm [16] reorders the rows and columns of the input dissimilarity matrix, $D \mapsto D^*$, and displays a grayscale image $I(D^*)$ whose ij-th element is a scaled dissimilarity value between x_i and x_j, and hence, objects o_i and o_j. Each element on the diagonal of the VAT image corresponds to zero distance. The reordered dissimilarity image $I(D^*)$ is often called a *cluster heat map*, very much in evidence in bioinformatics and neuroscience [17, 18]. Off the diagonal, the values range from 0 to 1. If an object is a member of a cluster, then it also should be part of a submatrix of "small" values, which appears as a darker block along the diagonal of the image matrix $I(D^*)$. The heuristic used with VAT/iVAT images is to estimate the number of clusters in the data by counting the number of dark submatrices that may occur along the diagonal.

VAT works like this: after building D from X, find the longest edge in the graph (the largest distance in D). Starting at either end, build a *minimal spanning tree*(MST) by inserting the next shortest safe edge into the current tree. The construction of the MST in VAT uses a slight variation of Prim's algorithm [19]. When the MST is completed, construct D* by reordering the rows and columns of D using the order of insertion of edges into the MST.

An improved version of VAT called iVAT [20] transforms $D \mapsto D'$ using a path-based distance and then VAT is applied to D' to get D'^*, resulting in an iVAT image $I(D'^*)$. This is a feature extraction technique applied to D. The recursive computation of D'^* given in [20] is $O(n^2)$. Appendix 1 contains the pseudocode for both VAT and iVAT. The Matlab code for them is readily available by contacting the authors.

VAT and iVAT are quite user-friendly in the sense that there are no parameters to pick other than the dissimilarity function (d) used to convert X to D. This input matrix can actually be a bit more general than a true distance because its only requirements are metric properties M1 and M2: $D = D^T; d_{ij} \geq 0 \forall i,j; d_{ii} = 0 \forall i$. The most important point about this display technique is that it is applied directly to (the distance matrix of) the upspace data, so there is no distortion of the structural information introduced by a feature extraction function from the upspace to a chosen downspace.

To illustrate, the data in Fig. 2a appear to have c = 3 visually apparent clusters of concentric points in the plane. These 2D vectors are converted to pairwise Euclidean distances to obtain the dissimilarity data matrix $D = [d_{ij}] = [(x_i - x_j)^T(x_i - x_j)]$. The VAT image $I(D^*)$ of the VAT reordered data, $D \mapsto D *$, appears in Fig. 2b. Do you see any hint of the structure of the data in this view? No (Fig. 2).

This happens because there are in this data set a few inliers between the three rings, and when VAT forms the MST of this data, it jumps back and forth from one ring to another, instead of calmly collecting all of the edges in each ring before moving to the next one. So, even with rearrangement by MST ordering, the VAT image is quite deceiving—it suggests that the input data has very little cluster structure, when in fact it has quite a lot.

Applying iVAT to D results in the iVAT image $I(D'^*)$ in Fig. 2c. The three dark blocks along the diagonal of $I(D'^*)$ strongly suggest that the input data may have c = 3 clusters. Another point: the size of the three diagonal blocks corresponds roughly

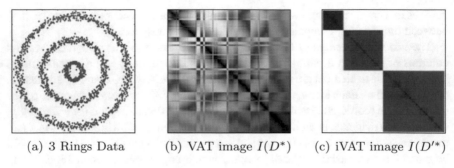

(a) 3 Rings Data   (b) VAT image $I(D^*)$ (c) iVAT image $I(D'^*)$

Fig. 2 Visual assessment of clustering tendency with VAT and iVAT images

to the number of points in each ring. The smallest, uppermost diagonal block in Fig. 2c represents the smallest inner cluster in Fig. 2a. The larger center diagonal block represents the larger cluster surrounding the inner points. And the largest block corresponds to the outer cluster in the data.

It is very important to understand that an iVAT image merely suggests that the input data are structured this way: iVAT does not find any partition of the object data in Fig. 2a. Since we can see the apparent cluster structure in this input data, this visual assessment of clustering tendency seems unnecessary. But it's quite simple to make a data set just like this in, say, 4-dimensional space, and then we are out of luck because no visual assessment by humans is possible. When this happens, however, VAT and iVAT can often still "see" cluster structure in the high dimensional data. Since VAT and iVAT can produce images such as these from data of arbitrary dimensions, we can use them to make a visual estimate of possible cluster structure in the upspace. They can also be wrong, but these images (especially the iVAT image) usually give us at least some idea about what kind of cluster structure the input data might contain; that is, they provide us with clues about potential starting points for Problem P2: finding a useful partition of the input data. Next, we discuss the connection of VAT and iVAT to Dunn's index and single linkage clustering.

5 Single Linkage Clustering, iVAT and Dunn's Index

Agglomerative single linkage clustering works like this: Begin with all n objects (the n vectors in X) as singleton clusters. A function d that measures pairwise dissimilarities on $X \times X$ is chosen, and used to build an $n \times n$ dissimilarity matrix $D_n = [d(\mathbf{x}_i, \mathbf{x}_j)]$. Assuming there are no distance ties (if there are, resolution can take several forms), the closest pair of vectors, say $\mathbf{x}_i$ and $\mathbf{x}_j$, are merged to form the cluster $\{\mathbf{x}_i, \mathbf{x}_j\} = \mathbf{x}_i \cup \mathbf{x}_j$. Now the SL set distance (Eq. 3) between $\mathbf{x}_i \cup \mathbf{x}_j$ and the $n - 2$ remaining singleton clusters is computed, the ith and jth rows in D_n are deleted, and a new row and column (the $n - 1$ set distances) is added, resulting in the reduces dissimilarity matrix D_{n-1} of size $(n - 1) \times (n - 1)$. This procedure continues until it terminates with all n vectors in the same cluster. The result is a hierarchical set of nested clusters beginning at $c = n$ and terminating at $c = 1$. In other words, if there are no ties, SL generates a unique partition U_k of X for each $c = k, k = n$ to 1.

Gower and Ross [21] were perhaps the first authors to point out that all the information needed to build a SL hierarchy is contained in a minimal spanning tree of the data. They showed how to recover the SL clusters produced by applying the above procedure to D_n by instead constructing the MST of D_n (i.e., of the underlying graph G), followed by making a backpass through the MST, cutting the tree at each minimum weight edge during this pass, into subtrees.

After the MST is built, the backpass through the tree cuts the sequence of connecting edges in descending order of magnitude. The first cut will occur at the maximum SL merger distance, and thence to the second largest, third largest, and so on, until the tree is cut into n leaves. The vertices in the subtrees at any level are the SL clusters at that level. Appendix 2 contains pseudocode for SL which indicates the role of the MST in the forward (building the MST) and backward (cutting the MST to get the clusters) passes.

Recall that the order of edge insertion in this same MST is used by iVAT to reorder the rows and columns of D_n to produce an iVAT image of the input data. Thus, there is an intimate connection between SL clustering and iVAT. Moreover, both methods are tied to Dunn's index at (4) through the use of Eq. (3), which computes SL distances with respect to a chosen dissimilarity function d. So, there is a natural connection between the three algorithms: Dunn's index, iVAT, and SL clustering. Figure 3a–d excerpted from ([22], Fig. 2), illustrate this relationship.

The view in Fig. 3a is a scatterplot of X, 100 points in 2-space. It is not immediately clear (to human observers) how many clusters this data set contains (have a crack at it before looking at the remaining views). Euclidean distances are computed on pairs of input data, resulting in the 100×100 Euclidean distance matrix D_{ED}. Figure 3b is the iVAT image of D_{ED}. There are $c = 10$ darkest blocks in this image, but there are also other attractive suggestions for this visual assessment. For example, there are apparently $c = 4$ dominant blocks in this image that contain the $c = 10$ smaller and darker blocks along the diagonal. An attractive feature of iVAT images is that they can suggest cluster structure at more than one value of c. Figure 3c shows the MST of D_{ED} (the green edges that link the 10 subsets). And finally, Fig. 3d is the SL partition of the 100 points obtained by the backpass through the MST that cuts the 9 largest edges.

Dunn's index with d = ED for the partition shown in Fig. 3d is 0.26, so this is not a CS partition in the sense of Eq. (5). Havens et al. studied the relationship between VAT (and hence iVAT) and single linkage in [23]. They presented an analytical comparison of the two algorithms in conjunction with numerical examples to show that VAT reordering of dissimilarity data is directly related to the clusters produced by single linkage hierarchical clustering. The preservation of VAT reordering passes to iVAT, and this fact underlies the extension of SL and iVAT to the big data case [22, 24].

In [22] it was established that clusiVAT produces true SL clusters in data that are CS in the sense of Dunn. This method was extended to the general case of clustering

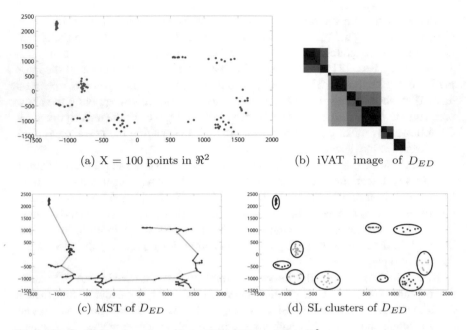

(a) X = 100 points in $\Re^2$ (b) iVAT image of D_{ED}

(c) MST of D_{ED} (d) SL clusters of D_{ED}

Fig. 3 Single linkage clustering with the MST of 100 points in $\Re^2$

in big data that are not necessarily CS, and compared to three k-means based methods as well as Cure in [22]. The relationship between Dunn's index and contrast in VAT images was discussed in [25], where an analytical comparison in conjunction with numerical examples demonstrated that the effectiveness of VAT in showing cluster tendency is directly related to Dunn's index.

To summarize, we know that Dunn's index is a natural companion to iVAT images and that iVAT is explicitly linked to single linkage clustering via their use of the same MST. In the next section we will explore the use of Dunn's index in conjunction with iVAT images of waveform data.

6 Experiments with Simulated Waveform Data

Recording of extracellular action potential (spike) waveforms generated by neuronal activity is an important technique in neuroscientific research. However, distinguishing the number of units present within recordings from a single electrode remains a fundamental and important technical issue. The first step in identifying distinct units begins by clustering unlabelled spike trains with any one of a number of unsupervised learning algorithms such as hard c-means, fuzzy c-means, single linkage, or Gaussian mixture decomposition. One aim of the present work is to improve the reliability of single unit identification with the use of iVAT for visual assessment of

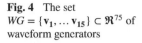

Fig. 4 The set
$WG = \{\mathbf{v_1}, \ldots \mathbf{v_{15}}\} \subset \mathfrak{R}^{75}$ of
waveform generators

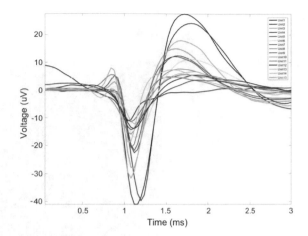

potential cluster structure in the data. This begins by showing the intrinsic relationship between the quality of iVAT images and Dunn's index. We will demonstrate this using simulated waveforms that are surrogates for spike trains.

Figure 4 is a set of 15 graphs of (simulated) waveforms that we use as waveform generators. Each generator is a vector of 75 voltage amplitudes that are sampled at equal time intervals along the horizontal axis. We denote the set of 15 exemplars as $WG = \{\mathbf{v_1}, \ldots \mathbf{v_{15}}\} \subset \mathfrak{R}^{75}$. The waveform generators in Fig. 4 and their corresponding set of waveforms are taken from dataset made available online by Regalia et al. [26], which were built as follows. The waveform generators were built by averaging the spontaneous neuronal spike activity from brain cultures which were classified and labeled by expert human operators. The waveform $\mathbf{v_j} \in WG$ is used to produce a crisp set of n_j waveforms, $X_j = \{\mathbf{x_{1j}}, \ldots \mathbf{x_{n_j j}}\} \subset \mathfrak{R}^{75}$, by first choosing an integer n_j randomly from the set $\{600, \ldots, 1000\}$. Then the background noise, taken from sections of the recorded signals that didn't contain spikes, was normalized and overlapped with the average waveform with selected signal to noise ratios to make n_j waveforms. The set of n_j waveforms mimics the average in vivo firing rates of between 3 and 8 Hz in a 60 s recording. Finally, recordings from a group of units were simulated by appending waveforms of 2 or 3 of the sets. The order of waveforms was uniformly randomized to mimic the random occurrence of spikes in recordings.

While these 15 waveforms share similar profiles, it is clear that there are differences across the set that have been shown to correlate with physiologically different neurons [27]. If asked to choose the most anomalous waveforms in this set, most observers would identify units 2 (red) and 4 (black) as being "outliers" to the general trend of the other 13 graphs due to their larger amplitudes and displacement to the right of the overall "centers" of the set of graphs. We will see that these two waveforms are recognized as anomalous by iVAT and single linkage.

Figure 5a is the iVAT image of these 15 vectors made with an input distance matrix with pairwise Euclidean distance values, denoted here as d = ED, in the 75 dimensional upspace. The integers aligned along the horizontal and vertical borders

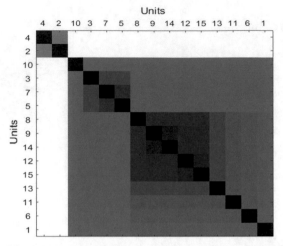

(a) iVAT image of the upspace data set WG with d=ED

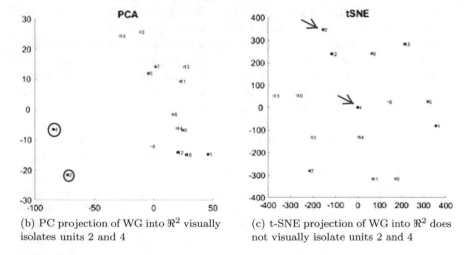

(b) PC projection of WG into $\Re^2$ visually isolates units 2 and 4

(c) t-SNE projection of WG into $\Re^2$ does not visually isolate units 2 and 4

Fig. 5 iVAT image of WD in $\Re^{75}$ and two projections of WG in $\Re^2$

of the image are the indices of the 15 signals after reordering of the input data. Waveforms 2 and 4 are isolated in Fig. 5a by the 2×2 darker block in the upper left corner of the image. The other 13 waveforms are grouped together (by iVAT reordering) and represented by the 13×13 block at the lower right. This highlights the utility of iVAT imagery to detect anomalies in high dimensional data upspaces, a fact that is well documented for trajectory data (which is a special type of waveform data) in [28].

Figure 5b is a scatterplot of the first two *principal components* (PC, [2]) of the set WG. This is a projection of the upspace data in $\Re^{75}$ into the PC downspace $\Re^2$.

Fig. 6 The three waveshape generators v_2, v_8, v_{13}

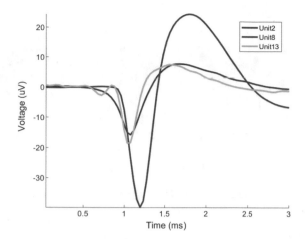

The vectors labelled 2 and 4 which are circled in this view are somewhat isolated from the other 13. This further supports our assertion that units 2 and 4 are visually different with respect to the other 13 wave generators. However, note that the t-SNE projection [29] of these 15 waveforms in Fig. 5c produces a scatter plot that does *not* isolate units 2 and 4 (red arrows) at all. This exemplifies one of the vagaries of visual assessment (Plato's cave) based on downspace projections: they often present very different interpretations of the upspace data.

We used the 15 vectors in WG to generate 15 clusters of waveforms in the manner described above. These 15 sets become the basis for constructing (simulated) sets of clusters for c = 2, 3, etc. For example, we can build 105 sets corresponding to two clusters as $X_{c=2} = \{X_{ij} = X_i \cup X_j : 1 \leq i \leq 15; i+1 \leq j \leq 15\}$. A union of any three distinct sets gives a set X_{ijk} of c = 3 clusters; and so on. Built this way, we have labelled data with corresponding ground truth partitions that enable us to compute the Dunn's index for any juxtaposition of subsets of signals. For example, the ground truth partition for the simulated clusters $X_{2,13} = X_2 \cup X_{13}$ is a $2 \times (n_2 + n_{13})$ matrix $U \in M_{h2(n_2+n_{13})}$ which looks like this:

$$U_{2,13} = \left[\begin{bmatrix} 1 \ 1 \ \dots \ 1 \\ 0 \ 0 \ \dots \ 0 \end{bmatrix}_{2 \times n_2} \begin{bmatrix} 0 \ 0 \ \dots \ 0 \\ 1 \ 1 \ \dots \ 1 \end{bmatrix}_{2 \times n_{13}} \right] \qquad (6)$$

In the interest of brevity, we discuss just three of the cases for c = 2 clusters in this article. Specifically, we will discuss experiments using units 2, 8 and 13, which are shown in Fig. 6. We envision the outcome of this study as the basis for a much more detailed extension of these experiments using other simulated sets of waveform clusters in a future article.

You can see in Fig. 6 that the blue outlier signal (v_2) is quite different from v_8 and v_{13}. On the other hand, wave generator units 8 and 13 appear to be very similar to each other in both amplitude and shape, so our expectation is that the crisp clusters

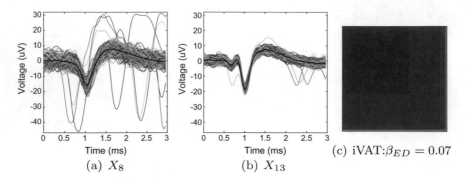

Fig. 7 Clusters generated by **a** v_8 **b** v_{13}, and **c** iVAT image of $X_{8,13}$

comprising $X_{8,13}$ may not be very separable (i.e., $X_{8,13}$ is likely to be interpreted as having a single cluster by iVAT. The Dunn's index based on the $c = 2$ ground truth partition of $X_{8,13}$ should be quite small.). Our analysis will be confined to the three 2 cluster data sets $X_{8,13}$, $X_{2,8}$ and $X_{2,13}$.

The two first graphs (a and b) in Figs. 7, 8 and 9 show all the waveforms of each cluster (different clusters) with the generating prototype shown as the heavy black waveform. These figures also show the iVAT and Dunn's index outputs for these three data sets noted as β_{ED} since the distance used in all three figures is Euclidean distance.

As expected, the clusters corresponding the sets of waveforms in Fig. 7 have a very low $\beta_{ED} = 0.07$ for data set $X_{8,13}$, and the iVAT image of this data set suggests that there is only one cluster in the data. Our supposition is that this happens because the generators v_8 and v_{13} are (visually) quite similar. Note that the waveforms in cluster X_8 include 6 or 7 that are wildly dissimilar from the generator of this cluster, and these anomalies probably cause the MST built by iVAT to jump back and forth between X_8 and X_{13}, so the MST of the data set $X_{8,13}$ will not isolate two clusters, and indeed, we feel that it should not. Instead, all of the vectors are joined in one large cluster under iVAT reordering.

The blocks in the iVAT image for data set $X_{2,8}$ in Fig. 8 are not sharply delineated, but they are certainly visually apparent, suggesting that there are $c = 2$ clusters in this data set. Dunn's index for $X_{2,8}$ is $\beta_{ED} = 0.25$, a moderate increase from that for $X_{8,13}$.

The iVAT image for data set $X_{2,13}$ in Fig. 9 is very clear and distinct. Notice how clean and distinct the data set X_2 is: all of the generated waveforms in this cluster adhere closely to the shape of the generating prototype. Figure 9c clearly suggests that $X_{2,13}$ contains $c = 2$ clusters, and it's Dunn's index $\beta_{ED} = 1.13$ tells us that this data set is compact and separated in the sense of Eq. (5) with respect to d = ED. As an aside, we computed Dunn's index on this data set using the 1-norm (the city block or Manhattan metric, d = D1) and the sup-norm (or Chebyshev d = Max) metrics: $\beta_{D1} = 1.06 > 1$ and $\beta_{Max} = 1.14 > 1$ This shows that $X_{2,13}$ is CS in the sense of Eq. 5

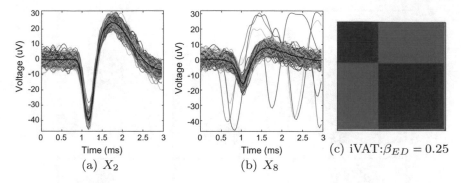

(c) iVAT:$\beta_{ED} = 0.25$

(a) X_2 (b) X_8

Fig. 8 Clusters generated by **a** $\mathbf{v_2}$ **b** $\mathbf{v_8}$, and **c** iVAT image of $X_{2,8}$

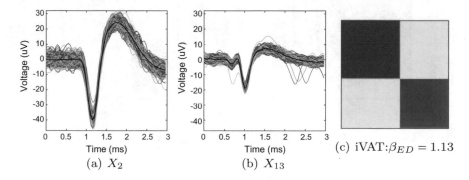

(c) iVAT:$\beta_{ED} = 1.13$

(a) X_2 (b) X_{13}

Fig. 9 Clusters generated by **a** $\mathbf{v_2}$ **b** $\mathbf{v_{13}}$, and **c** iVAT image of $X_{2,13}$

for any of these three choices of d, all of which are bona fide metrics on the input space.

Another way to get some insight about the results in Figs. 7, 8 and 9 is to examine the visual suggestions provided by projections into a 2D downspace. Figure 10 shows the t-SNE [29] projections of the three data sets $X_{8,13}$, $X_{2,8}$ and $X_{2,13}$ with (upper row) and without (lower row) labels (the scatter plots with colors in the upper row indicate the ground truth tags of each data point).

These scatter plots are consistent with the inferences drawn about the three data sets made by visual inspection of upspace iVAT images. In particular, we draw your attention to the lower view in Fig. 10 for the unlabeled t-SNE view of $X_{8,13}$, which agrees with the visual assessment of $X_{8,13}$ provided by Fig. 7c which indicates that there is only one cluster of waveforms in this data set. In our expanded study of these experiments we will see several cases where there is disagreement between the upspace iVAT and downspace t-SNE and PC visualizations.

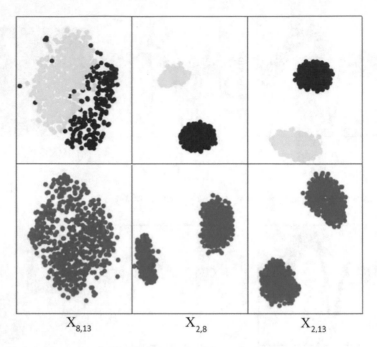

$$X_{8,13} \qquad\qquad X_{2,8} \qquad\qquad X_{2,13}$$

Fig. 10 Labeled (upper) and Unlabeled (Lower) t-SNE projections

7 Customizing Dunn's Index for Waveform Data

Our notation for Dunn's index at (4) was chosen to emphasize that it can be computed using any metric d on a real vector space that satisfies the metric properties M1-M3. The iVAT algorithm is even less demanding, as it only requires properties M1 and M2. Both of these algorithms, and SL as well, are perfectly happy to operate on any square dissimilarity matrix that satisfies M1 and M2. In most clustering problems involving feature vector data, the data do not dictate a particular choice for d. However, the data we are interested in always comes to us in the form of finite time series (waveforms) represented by vectors of amplitudes sampled at pre-specified time intervals. The literature is replete with alternatives that claim to be superior to the default choice of Euclidean distance $\beta_D(U|d = ED)$. In this section we explore the possibility that choosing a "wave-specific" measure of dissimilarity to compute the input matrix D from X leads to a different, possibly better, iVAT image coupled with more robust values of Dunn's index. The customized Dunn's indices we will study in this section are:

$\beta_D(U|d = ED)$ Euclidean distance   (7a)

$\beta_D(U|d = CC)$ One minus the normalized correlation   (7b)

$\beta_D(U|d = DTW)$ Dynamic Time Warping   (7c)

$\beta_D(U|d = SBD)$ The shape based distance (SBD) in Appendix 3   (7d)

$\beta_D(U|d = \sqrt{SBD})$ The square root of the SBD   (7e)

Many authors refer to all of these functions as distances, a term that carries the implication that they are genuine metrics. Certainly, ED is a metric, the usual choice for DI, as illustrated in Fig. 1 which has also been commonly employed as a distance measure between neuronal spike waveforms [30]. DTW and cross-correlation are not metrics [31, 32]. Nevertheless, both have been successfully used as "similarity" measures for waveform data. DTW has long been used in speech recognition applicaitons as well as recent applications in neuroscience [33]; cross-correlation is widely used in template matching techniques in neuroscience [34]. It is worth displaying the CC measure we use above in 7b; for two vectors $\mathbf{x}$ and $\mathbf{y}$:

$$CC(\mathbf{x}, \mathbf{y}) = 1 - \frac{(\mathbf{x} - \bar{\mathbf{x}})^T (\mathbf{y} - \bar{\mathbf{y}})}{\|\mathbf{x} - \bar{\mathbf{x}}\| \|\mathbf{y} - \bar{\mathbf{y}}\|} \qquad (8)$$

If $\mathbf{x}$ and $\mathbf{y}$ are standardized vectors (i.e. standardized to z-scores with zero mean and standard deviation of 1), there is a direct relationship between the CC and the Euclidean distance according to the cosine theorem: $ED = \sqrt{2(p-1)CC}$ when p is the dimension of the vectors. Therefore, for the special case of standardized vectors, the CC measure is a metric (or better said: it is the squared Euclidean distance with a constant multiplier). In general, however, $CC(\mathbf{x}, \mathbf{ax}) = 0$ for any $a > 0$, but $\mathbf{x} \neq \mathbf{ax}$, so CC is not a metric on $\mathfrak{R}^p \times \mathfrak{R}^p$. This scale-free property of cross-correlation, desirable in some similarity evaluation scenarios, is what prevents it from being a distance metric.

As mentioned above, cross-correlation has been employed as a matching index between neural waveform data and a template waveform [34, 35]. Paparrizos et al. [13] used one minus maximum normalized cross-correlation as the dissimilarity measure between any two waveform time series and they introduce the name Shape based distance (SBD) for it. We use their algorithm (cf. Appendix 3) and borrow their nicely put name for it in this work. It is worth noting that SBD is essentially the CC between one waveform and a shifted version of the other waveform. The shift is calculated to produce the maximum correlation between the waveforms. The first step of the algorithm is to standardize the waveforms. We also computed the square root of the SBD values to see if the behaviour was different.

We call the argument (d) in the last four functions at (7) dissimilarity measures. They can all be used to build an input dissimilarity matrix D for input to iVAT and SL and they can all be substituted into Dunn's index. We will call the indices shown in (7b-7e) as *customized Dunn's indices* (cDIs).

Here is a summary of our experimental protocols:

1. Choose a data set X_{ij} with ground truth partition U_{ij}
2. Choose a dissimilarity measure d from the choices shown in Eq. 7
3. Compute the DI (d = ED) or cDI on U_{ij} using d and X_{ij}
4. Build an input matrix $D = [d(x, y) : x, y \text{ in} X_{ij}]$
5. Get iVAT(D)
6. Run SL(D)
7. Discuss results

Table 1 compares Dunn's indices on the ground truth partitions of the three test data sets $X_{8,13}$, $X_{2,8}$ and $X_{2,13}$ as described in Sect. 6 based on building the input matrix D for iVAT and SL with the five dissimilarity measures (d) in (7). We have repeated the values shown for $\beta_D(U|d = ED)$ in Sect. 6 as the first row in Table 1 to make comparisons easier. The most striking feature of this table, seen graphically in Fig. 11, is that the first three of the five distances, ED, CC, and DTW all rank the three cases in ascending order of Dunn's index, while the two shape based distances produced using Algorithm 1 in [13] listed as Appendix 3 show the maximum value of Dunn's index for the data set $X_{8,13}$.

The graphs in Fig. 11 show that the trend of Dunn's index for ED, DTW and CC are very similar. The graphs for SBD and its square root are similar to each other, but quite different from the other three. We can make a finer distinction by examining the iVAT images for the five measures on all three of our test data sets.

Table 1 Values of customized Dunn's indices (cDI) on $X_{8,13}$, $X_{2,8}$ and $X_{2,13}$

Distance	$X_{8,13}$	$X_{2,8}$	$X_{2,13}$
$d = ED$	0.071	0.253	1.133
$d = CC$	0.036	0.053	0.331
$d = DTW$	0.049	0.119	0.793
$d = SBD$	0.076	0.035	0.050
$d = \sqrt{SBD}$	0.276	0.186	0.223

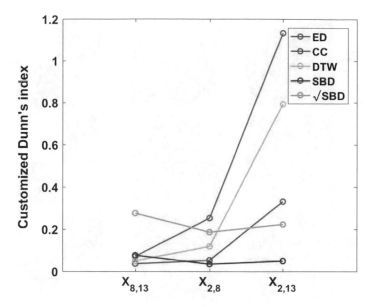

Fig. 11 cDIs on $X_{8,13}$, $X_{2,8}$ and $X_{2,13}$ using 5 dissimilarity functions

We discuss the iVAT images in Fig. 12 column by column. Data set $X_{8,13}$: Recall from Fig. 6 that graphs of the wave generators $\mathbf{v_8}$ and $\mathbf{v_{13}}$ are quite similar, and therefore, the clusters in X_8 and X_{13} are not expected to be very well separated. The iVAT images built with ED, CC and DTW are visually quite similar, and suggest that all of the points in $X_{8,13}$ are in a single cluster. This is consistent with our expectations. Moreover, the values of Dunn's indices for these three experiments are quite low, roughly in the range 0.03–0.07. The two SBD based images for $X_{8,13}$ in Fig. 12 are visually similar to each other, but somewhat different than the upper three panels. These two images have an indistinct but visible darker block in their upper left portions, fading into a bland area of almost uniformly grey values over the remainder of the image. The visual suggestion is that there is only one cluster, but it has a primary core that is tight enough to be seen embedded in the overall grey background. Dunn's index for SBD is very similar to the first three values: the square root of SBD is much larger.

Data set $X_{2,8}$: The center column of Fig. 6 is the intermediate case. Here the functions d = ED and d = DTW produce iVAT images that show definite cluster structure at c = 2, whereas d = CC continues to produce an image that supports a belief in just one cluster. A closer look reveals a thin grey border at the top and right in all three of these images. Our supposition is that these pixels correspond to the 5 or 6 spurious waveforms that are visually evident in Fig. 8b (or Fig. 7a); meaning that

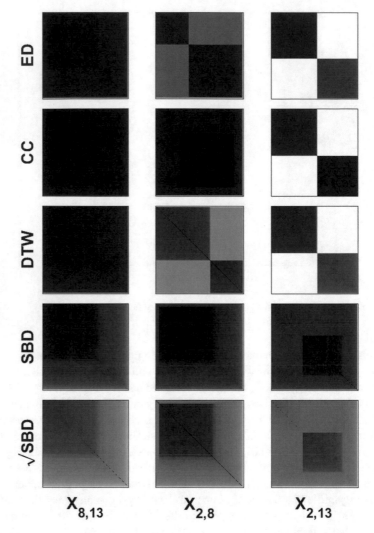

Fig. 12 iVAT images of $X_{8,13}$, $X_{2,8}$ and $X_{2,13}$ based on input dissimilarity matrix D built with 5 different dissimilarity measures

they are really noise. The two SBD based images for this data set are surprisingly similar to the ones for data set $X_{8,13}$ in their left. A bit of a mystery; the implication is that the SBD distances don't make as sharp a distinction between clusters as the other three dissimilarity measures. Table 1 values show a fairly sizeable change of about 0.18 for d = ED, and a less dramatic change of about 0.07 for DTW. The CC value, however, is only increases by 0.017, a really small jump, and its iVAT image reflects this, basically, no change in the image or the Dunn's index value. In contrast, the two SBD indices in Table 1 decrease to their minimums in the three experiments.

Data set $X_{2,13}$: The right column of Fig. 6 is the most interesting and distinctive case. Here the three functions d = ED, d = CC and d = DTW all produce nice crisp visualizations that correspond to c = 2 clusters. And the Dunn's index values for these three cases in Table 1 show much sharper jumps: roughly 0.88, 0.28 and 0.68 for ED, CC and DTW. Consulting Eq. (5) shows that $X_{2,13}$ is compact and separable with respect to the Euclidean metric in the sense of Dunn's theorem. On the other hand, the images built with the two SBD distances continue to confound: the visually darker sub-block in these two panels seems to have floated out into the middle, embedded in a sea of noise. Table 1 shows little change in the cDIs for these two cases.

Finally, please correlate each row of customized Dunn's indices in Table 1 to the corresponding row in Fig. 12. The visual acuity of the images in each row is clearly related to values of the cDIs: higher values of cDI indicate a clearer iVAT visualization. And conversely, when the cDIs are relatively flat and low, as they are for the two SBD measures, the images don't change much and are not very informative either.

8   Single Linkage Revisited

Experiments in this section explore the relationship between values of Dunn's index and single linkage clusters derived from the underlying metric in use. Let $U_{GT} \in M_{hcn}$ be a ground truth partition for any labelled data set (X or R), and let $U \in M_{hcn}$ be a c-partition of it found by processing the dataset with any clustering algorithm. The *partition accuracy* of U with respect to U_{GT} is

$$PA(U|U_{GT}) = \frac{\#matched}{\#tried} \qquad (9)$$

We ran the single linkage algorithm using the five different dissimilarity measures in (7) on the three datasets $X_{8,13}$, $X_{2,8}$ and $X_{2,13}$ and got the crisp partitions for c = 2. Table 2 exhibits the values of the partition accuracy for each of the 5 measures on each of the three data sets. The first three rows of Table 2 confirm that the three cases, $X_{8,13}$, $X_{2,8}$ and $X_{2,13}$, have increasingly well separated clusters, manifested by the values increasing from about 0.35 ($\approx 1/3$ correct) to 0.61 ($\approx 2/3$ correct) to 1.00 (exact). This is pretty good agreement with the left to right sequences of images in Fig. 12, and the left to right values in Table 1 for the Dunn's indices. The SL partitions obtained using the first three dissimilarity measures (ED, CC, DTW) were all very similar for all three data sets. The partitions obtained by using d = SBD and its square root to define the SL distance at (3) were similar to each other, and produced comparable partition accuracies to the first three measures for $X_{8,13}$, $X_{2,8}$, but oddly these two measures produced very poor partition accuracy for $X_{2,13}$, the (apparently) most separable pair of clusters for the three data sets.

Perhaps it is not so odd when you look at the results from this angle: for the first two datasets $X_{8,13}$ and $X_{2,8}$, all the measures give relatively poor results. Analyzing the results of the SL clustering by examining the actual partitions for these two datasets, SL has clustered all but one or two points into one big cluster, i.e. We get two clusters by SL but one only has one or two of the points and the rest are grouped together as in a very large cluster. The result is a low match between the ground truth and SL labels.

As mentioned above (cf. Table 2) the SBD measure didn't facilitate a good SL clustering for the $X_{2,13}$ case either. Figure 13 shows the dissimilarity measure between each of the two waveforms of the $X_{2,13}$ by the two measures: SBD and CC. Unit2 has 292 waveforms, and Unit13 has 343 waveforms, therefore the $X_{2,13}$ set has 635 waveforms. The total number of distances between pairs is 201,295 seen along the horizontal axis of Fig. 13. The y-axis represents the value of dissimilarity given by each of the measures on the 201,295 pairs. You can see that the CC (red curve) has two phases or steps below and above the value 0.3, but the SBD measure has a

Table 2 Partition accuracy for crisp SL partitions at c = 2

Partition accuracy	$X_{8,13}$	$X_{2,8}$	$X_{2,13}$
$d = ED$	0.352	0.610	1.000
$d = CC$	0.352	0.610	1.000
$d = DTW$	0.348	0.615	1.000
$d = SBD$	0.350	0.613	0.463
$d = \sqrt{SBD}$	0.350	0.613	0.463

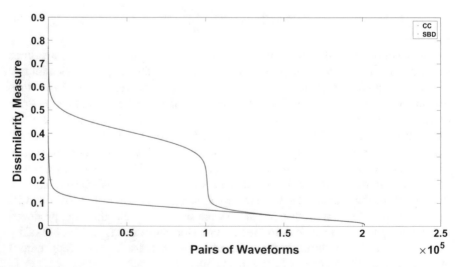

Fig. 13 The dissimilarity measures CC (blue) and SBD (red) between each pair of waveforms in $X_{2,13}$

gradual decrease. This shows why the CC values help i VAT form distinct blocks and why SL is able to distinguish two clusters. Whereas this is not achievable using the SBD measure. We believe this is caused by standardization (z normalization) of the waveforms in the first step of the SBD algorithm. This normalization removes the scale difference between the waveforms in X_2 and X_3 that contributes to differentiation between them with the other measures. Whether standardization is appropriate for neuronal spike waveforms recorded by an extracellular electrode is left an in intriguing question that will be discussed in our future paper.

9 Conclusions

The values in Tables 1 and 2 show that the larger the cDI, the better will be the match of the SL clusters to the ground truth. Correlating this with the i VAT images shown in Fig. 12 establishes a clear link between the three tools: as the cDI increases, the i VAT image of the upspace data exhibits better visual acuity, and the corresponding SL partition of the upspace data secures better agreement with the ground truth labels.

While this small set of experiments hardly provides enough evidence to support any statistically valid conclusions, it does suggest that Dunn's index using Euclidean distance is a fairly reliable indicator of how much faith we should place on interpretations of cluster structure in upspace data based on i VAT images of that data. We expect larger values of $\beta_D(U|d = ED)$ to correspond to clearer i VAT images, and conversely, values of $\beta_D(U|d = ED)$ near zero will usually indicate that the corresponding i VAT image will suggest there is only a single cluster in the upspace data, even if downspace projections suggest otherwise. Somewhat surprisingly, we did not find that customized versions of Dunn's index using any of the four waveshape specific measures of dissimilarity at (7b–7e) provided results that were superior to simple Euclidean distance. If the datasets had waveforms whose peaks and valleys occurred at different time intervals, perhaps one of these measures would be more effective than Euclidean distance.

It is tempting, but probably overoptimistic, to think that we can identify a threshold that reliably separates the good and bad cases based on values of Dunn's index. All of these methods can fail, and all of them do fail, so the best we can probably hope for is a "rule of thumb" heuristic that ties good i VAT images to larger values of Dunn's index. This will be a major focus of our next foray into the world of waveform clustering. The failure of Dunn's index in the presence of inliers and outliers is well known, so we are also anxious to insert these alternative waveform-based measures of dissimilarity into the generalized Dunn's indices discussed in [12].

We know that $CC(\mathbf{x}, \mathbf{y})$ and $SBD(\mathbf{x}, \mathbf{y})$ fail property M1 and hence are not true metrics on $\mathfrak{R}^p$. Lacking this, we can be sure that Dunn's theorem will not be verifiable using d = SBD. Moreover, it is not clear whether a value emerging from Algorithm 1 should be regarded as a squared distance or not, although the authors state that it replaces $\|\mathbf{x} - \mathbf{v}\|^2$ in the k-means problem. Finally, Eq. (13) in [13], and the text describing it, is questionable. The authors don't actually solve (13) directly, but use the Rayleigh Quotient theorem. If (13) is solved directly, the cluster centers would never change because the optimization is only over the previous set of cluster centers. Also, when they convert to the Rayleigh Quotient, they change the vector meaning of the input vectors from row to column format. More tests of the SBD model of [13] are needed before it can be regarded as a useful alternative to simply using ED with Dunn's index and iVAT.

A provocative by-product of this article is that when a set of (crisp) candidate partitions is produced by single linkage clustering of any data set, our experiments show that Dunn's index (in any form) should occupy a pre-eminent spot in the set of crisp internal CVIs available to evaluate the candidates. Arbelaitz et al. [10] studied the efficacy of 30 internal CVIs to evaluate candidates produced by three crisp clustering algorithms: k-means, Ward's MSE method, and average linkage. Generalized versions of Dunn's index introduced in [12] consistently scored in the top 5 of the 30 CVIs on a wide variety of synthetic and real data sets in [10]. Our study suggests that when SL clustering is used to produce candidates, some form of Dunn's index should stand out amongst its competitors. We plan to make this the focus of a future study.

And finally, all of the ideas put forward in this article need to be tested in the"real" case: upspace clustering of real unlabeled electrical recordings from brain tissue, in-vitro, aimed at sorting neuronal spike trains into clusters of different functional types of neurons. We will devote a future study to ascertaining whether the use of iVAT, Dunn's index, and SL clustering is a useful combination of tools for this problem in the neuroscience community.

Appendix 1. The VAT and IVAT Reordering Algorithms

	Algorithm VAT [15]
1	**In** D, $n \times n$ matrix of dissimilarities: $D = D^T$; $d_{ij} \geq 0 \ \forall \ i,j$; $d_{ii} = 0 \ \forall i$
2 3 4 5	**Set** $K = \{1, 2, ..., n\}$: $I = J = \varnothing$: Select $(i, j) \in \arg\max\{D_{st} : s \in K, t \in K\}$ $P(1) = i$: $I = \{i\}$: $j = K - \{i\}$ % Initialize MST at either end of edge with largest weight in D
6 7 8	**For** m = 2,..., n **do**: select $(i, j) \in \arg\min\{D_{st} : s \in I, t \in J\}$ Select $(i, j) \in \arg\min\{D_{st} : s \in I, t \in J\}$ $P(m) = j$: $I = I \cup \{j\}$: $J = J - \{i\}$: $d_{m-1} = d_{ij}$
9 10	**For** $1 \leq i, j \leq n$ **do**: $[D^*]_{ij} = [D]_{P(i)P(j)}$
11 12	**Out** VAT reordered dissimilarities D*: arrays P, d % Create VAT RDI I(D*) using D*
	Algorithm iVAT [16]
13	**In** D* =VAT reordered dissimilarity matrix: $D'^* = [0]$
14	**For** k = 2 to n **do**:
15 16	$j = \arg\min\limits_{r=1,...k-1} \{D^*_{kr}\}$
17 18	$D'^*_{kc} = D^*_{kc}$; c = j $D'^*_{kc} = \max\{D^*_{kj}, D'^*_{jc}\}$; c = 1,É, k-1;c-j
19 20	**For** j = 2,En,EÈ< j: $D'^*_{ji} = D'^*_{ij}$
	Out iVAT Reordered dissimilarities D'* % Create iVAT RDI I(D'*) using D'*

A.1 The input matrix D for VAT in line 1 is positive definite and symmetric. Any distance matrix will be of this type, but there are a number of cases that don't satisfy these constraints. And the size of D can be an issue. This basic version is only useful for fairly small values of n (say, n 10,000 or so). Extensions to rectangular, asymmetric and big data inputs are covered in the notes and remarks for this chapter.

A.2 Prim's MST algorithm usually starts at either end (i.e., vertex) of a smallest weight edge. Initialization at line 3 starts at the opposite extreme - either end of a largest weight edge. This prevents VAT from a certain type of off-course deviation that is discussed in Bezdek and Hathaway (2002).

A.3 The argmax and argmin function calls in lines 3, 7 and 15 produce sets, not single values. For example, in A4.1 is the set of all ordered pairs (i, j) that have a maximum distance. In case of ties, use a vertex from either end of any one edge in the set.

Appendix 2. Basic Single Linkage Clustering Algorithm

	Basic Single Linkage Clustering Algorithm [1]
1	**Input :** An $n \times n$ dissimilarity matrix $D = D_n \in M_n^+$
2	*% User choice for model parameter (set distance)*
3	Set distance for finite crisp subsets A, B
4	*% Initialization*
5	Initial (singleton) clusters : $I = \emptyset : J = V = \{1, 2, ..., n\}$
6	Pick starting object (vertex) $m \in V : I = \{m\} : J = J - \{m\}$
7	*% Build MST*
8	*For k = 2 to n*
9	% Find a minimum feasible edge e_{ij}
10	% No tie-breaking strategy is employed
11	Select $(i, j) \in \underset{v_p \in I. v_q \in J}{\arg\min} \{d_{pq}\}$
12	$I = I \cup \{j\}; \ J = J - \{j\}; \ d_{SL,k} = d_{ij}$
13	*Next k*
14	*% Extract nested partitions by backpass cuts in MST*
15	*For k = 2 to n*
16	Remove edge link having distance $d_{SL,k}$
17	C_k = connected subtrees (clusters of vertices) in the MST
18	*Next k*
19	*Outputs*: Crisp clusters , $C_n, C_{n-1}, ..., C_2, C_1$
20	Crisp partitions $\{U_n, U_{n-1}, ..., U_2, U_1\} \subset M_{hcn}$
21	Merger distances $\{d_{SL,k} : k = 2, ... n\}$

Appendix 3. Shape Based Distance

XIII.   APPENDIX 3. SHAPE BASED DISTANCE (ALGORITHM 1, [25])

Input: Two z-normalized sequences $x, y \in \mathfrak{R}^p$

1 length $= 2^{\text{nextpower}2(2*\text{length}(x)-1)}$

2 $CC = IFFT\{FFT(x, length) * FFT(y, length)\}$

3 $NCC_c = CC / \| x \| \bullet \| y \|$

4 [value,index] = max(NCCc)

5 dist = 1 − value

6 shift = index − length(x)

7 **if** shift ≥ 0 **then**

8   y' = [zeros(1, shift), y(1 : end−shift)]

9 **else**

10   y' = [y(1 − shift : end), zeros(1,−shift)]

11 **Output:** Dissimilarity dist = SBD(x, y) $\in \mathfrak{R}^+$;

         Aligned sequence **y'** of **y** towards **x**

References

1. Bezdek James C (2017) A primer on cluster analysis: 4 basic methods that (usually) work, 1st edn. Design Publishing, Sarasota, FL
2. Theodoridis S (2009) Pattern recognition. Academic Press, London. ISBN 978-1-59749-272-0
3. Duda RO, Hart PE, Stork DG (2012) Pattern classification. Wiley
4. Jain AK, Dubes RC (1988) Algorithms for clustering data. Prentice Hall College Div, Englewood Cliffs, NJ
5. Dubes R, Jain AK (1979) Validity studies in clustering methodologies. Pattern recognition, vol 11, no 4, pp 235–254, Jan 1979. ISSN 0031-3203. doi:10.1016/0031-3203(79)90034-7
6. Milligan GW, Cooper MC (1985) An examination of procedures for determining the number of clusters in a data set. Psychometrika, 50(2):159–179. ISSN 0033-3123, 1860-0980. doi:10.1007/BF02294245
7. Gurrutxaga I, Muguerza J, Arbelaitz O, Pérez JM, Martín JI (2011) Towards a standard methodology to evaluate internal cluster validity indices. Pattern Recognit. Lett., 32(3):505–515, February 2011. ISSN 0167-8655. doi:10.1016/j.patrec.2010.11.006
8. Dimitriadou E, Dolničar S, Weingessel A (2002) An examination of indexes for determining the number of clusters in binary data sets. Psychometrika, 67(1):137–159. ISSN 0033-3123, 1860-0980. doi:10.1007/BF02294713
9. Vinh NX, Epps J, Bailey J (2010) Information theoretic measures for clusterings comparison: variants, properties, normalization and correction for chance. J Mach Learn Res, 11:2837–2854. ISSN 1532-4435
10. Arbelaitz O, Gurrutxaga I, Muguerza J, Pérez JM, Perona I (2013) An extensive comparative study of cluster validity indices. Pattern Recognition 46(1):243–256. ISSN 0031-3203. doi:10.1016/j.patcog.2012.07.021
11. Dunn JC (1973) A fuzzy relative of the ISODATA process and its use in detecting compact well-separated clusters. J Cybern 3(3):32–57. ISSN 0022-0280. doi:10.1080/01969727308546046
12. Bezdek JC, Pal NR (1998) Some new indexes of cluster validity. IEEE Trans Syst Man Cybern Part B (Cybern) 28(3):301–315. ISSN 1083-4419. doi:10.1109/3477.678624

13. Paparrizos J, Gravano L (2015) K-shape: efficient and accurate clustering of time series. In: Proceedings of the 2015 ACM SIGMOD international conference on management of data, SIGMOD '15, New York, NY, USA, 2015. ACM, pp 1855–1870. ISBN 978-1-4503-2758-9. doi:10.1145/2723372.2737793

14. Morris BT, Trivedi MM (2008) A survey of vision-based trajectory learning and analysis for surveillance. IEEE Trans. Circuits Syst Video Technol 18(8):1114–1127. ISSN 1051-8215. doi:10.1109/TCSVT.2008.927109

15. Valdés JJ, Alsulaiman FA, Saddik AEl (2016) Visualization of handwritten signatures based on haptic information. In: Abielmona R, Falcon R, Zincir-Heywood N, Abbass HA (eds) Recent advances in computational intelligence in defense and security, number 621 in studies in computational intelligence. Springer International Publishing, pp 277–307. ISBN 978-3-319-26448-6 978-3-319-26450-9. doi:10.1007/978-3-319-26450-9-11

16. Bezdek JC, Hathaway RJ (2002) VAT: a tool for visual assessment of (cluster) tendency. In: Proceedings of the 2002 international joint conference on neural networks, 2002. IJCNN '02, vol 3, pp 2225–2230. doi:10.1109/IJCNN.2002.1007487

17. John N. Weinstein. A Postgenomic Visual Icon. *Science*, 319(5871):1772–1773, March 2008. ISSN 0036-8075, 1095-9203. doi:10.1126/science.1151888

18. Wilkinson L, Friendly M (2009) The history of the cluster heat map. Am Stat 63(2):179–184. ISSN 0003-1305. doi:10.1198/tas.2009.0033

19. Prim RC (1957) Shortest connection networks and some generalizations. Bell Syst Tech J 36(6):1389–1401. ISSN 1538-7305. doi:10.1002/j.1538-7305.1957.tb01515.x

20. Havens TC, Bezdek JC (2012) An efficient formulation of the improved visual assessment of cluster tendency (iVAT) algorithm. IEEE Trans Knowl Data Eng 24(5):813–822. ISSN 1041-4347. doi:10.1109/TKDE.2011.33

21. Gower JC, Ross GJS (1969) Minimum spanning trees and single linkage cluster analysis. J R Stat Soc Ser C (Appl Stat), 18(1):54–64. ISSN 0035-9254. doi:10.2307/2346439

22. Kumar D, Bezdek JC, Palaniswami M, Rajasegarar S, Leckie C, Havens TC (2016) A hybrid approach to clustering in big data. IEEE Trans Cybern 46(10):2372–2385. ISSN 2168-2267. doi:10.1109/TCYB.2015.2477416

23. Havens TC, Bezdek JC, Keller JM, Popescu M, Huband JM (2009) Is VAT really single linkage in disguise? Ann Math Artif Intell 55(3–4):237. ISSN 1012–2443:1573–7470. doi:10.1007/s10472-009-9157-2

24. Havens TC, Bezdek JC, Palaniswami M (2013) Scalable single linkage hierarchical clustering for big data. In: 2013 IEEE eighth international conference on intelligent sensors, sensor networks and information processing, pp 396–401. doi:10.1109/ISSNIP.2013.6529823

25. Havens TC, Bezdek JC, Keller JM, Popescu M (2008) Dunn's cluster validity index as a contrast measure of VAT images. In: 2008 19th international conference on pattern recognition, pp 1–4. doi:10.1109/ICPR.2008.4761772

26. Regalia G, Coelli S, Biffi E, Ferrigno G, Pedrocchi A (2016) A framework for the comparative assessment of neuronal spike sorting algorithms towards more accurate off-line and on-line microelectrode arrays data analysis. Comput Intell Neurosci 2016:e8416237. ISSN 1687-5265. doi:10.1155/2016/8416237

27. Barthó P, Hirase H, Monconduit L, Zugaro M, Harris KD, Buzsáki G (2004) Characterization of neocortical principal cells and interneurons by network interactions and extracellular features. J Neurophys 92(1):600–608. ISSN 0022-3077. doi:10.1152/jn.01170.2003

28. Kumar D, Bezdek JC, Rajasegarar S, Leckie C, Palaniswami M (2017) A visual-numeric approach to clustering and anomaly detection for trajectory data. Vis Comput 33(3):265–281. ISSN 0178-2789, 1432-2315. doi:10.1007/s00371-015-1192-x

29. van der Maaten L, Hinton G (2008) Visualizing data using t-SNE. J Mach Learn Res 9(Nov):2579–2605. ISSN 1533-7928

30. Rutishauser U, Schuman EM, Mamelak AN (2006) Online detection and sorting of extracellularly recorded action potentials in human medial temporal lobe recordings, in vivo. J Neurosc Methods 154(1–2):204–224. ISSN 0165-0270. doi:10.1016/j.jneumeth.2005.12.033

31. Ruiz EV, Nolla FC, Segovia HR (1985) Is the DTW "distance" really a metric? An algorithm reducing the number of DTW comparisons in isolated word recognition. Speech Commun 4(4):333–344. ISSN 0167-6393. doi:10.1016/0167-6393(85)90058-5

32. Wachman G, Khardon R, Protopapas P, Alcock CR (2009) Kernels for periodic time series arising in astronomy. In: Machine learning and knowledge discovery in databases. Springer, Heidelberg, pp 489–505. doi:10.1007/978-3-642-04174-7-32

33. Cao Y, Rakhilin N, Gordon PH, Shen X, Kan EC (2016) A real-time spike classification method based on dynamic time warping for extracellular enteric neural recording with large waveform variability. J Neurosci Methods 261:97–109. ISSN 0165-0270. doi:10.1016/j.jneumeth.2015.12.006

34. Kim S, McNames J (2007) Automatic spike detection based on adaptive template matching for extracellular neural recordings. J Neurosci Methods 165(2):165–174. ISSN 0165-0270. doi:10.1016/j.jneumeth.2007.05.033

35. Franke F, Pröpper R, Alle H, Meier P, Geiger JRP, Obermayer K, Munk MHJ (2015) Spike sorting of synchronous spikes from local neuron ensembles. J Neurophysiol 114(4):2535–2549. ISSN 0022-3077, 1522-1598. doi:10.1152/jn.00993.2014

A Shared Encoder DNN for Integrated Recognition and Segmentation of Traffic Scenes

Malte Oeljeklaus, Frank Hoffmann and Torsten Bertram

Abstract Detection of traffic related objects in the vehicles surroundings is an important task for future automated cars. Visual object recognition and scene labeling from onboard cameras provides valuable information for the driving task. In computer vision, the task of generating meaningful image regions representing specific object categories such as cars or road area, is denoted as semantic segmentation. In contrast, scene recognition computes a global label that reflects the overall category of the scene. This contribution presents an efficient deep neural network (DNN) capable of solving both problems. The network topology avoids redundant computations, by employing a shared feature encoder stage combined with designated decoders for the two specific tasks. Additionally, element-wise weights in a novel Hadamard layer efficiently exploit spatial priors for the segmentation task. Traffic scene segmentation is examined in conjunction with road topology recognition based on the cityscapes dataset [2] augmented with manually labeled road topology data.

1 Introduction

Fast and reliable environment representations deduced from the perception of the vehicles environment are a fundamental requirement for future automated driving. Camera-based methods extract the traffic related information solely from the visual appearance of a scene. In this context, scene segmentation denotes the task of computing semantically meaningful image regions corresponding to traffic related objects, such as cars or road area. In contrast, the task of traffic scene recognition infers a single global category that characterizes the essence of the entire scene. Together, these complementary representations facilitate a comprehensive interpretation of the traffic scene. This paper presents an efficient method capable of solving both problems in an integrated manner. Segmentation classifies each pixel into nine-

M. Oeljeklaus (✉) · F. Hoffmann · T. Bertram
Institute of Control Theory and Systems Engineering, TU Dortmund
University, Otto-Hahn-Str. 8, 44227 Dortmundrt, Germany
e-mail: Malte.Oeljeklaus@tu-dortmund.de

© Springer International Publishing AG 2018
S. Mostaghim et al. (eds.), *Frontiers in Computational Intelligence*,
Studies in Computational Intelligence 739,
https://doi.org/10.1007/978-3-319-67789-7_7

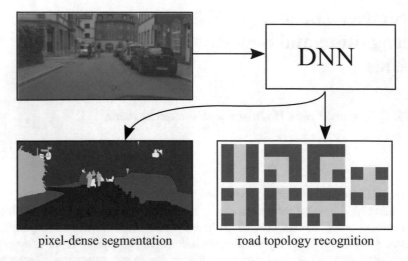

pixel-dense segmentation   road topology recognition

Fig. 1 Overview of the examined method

teen different object categories according to [2]. Traffic scene recognition is investigated w.r.t. to seven alternative road topologies as illustrated in the bottom right part of Fig. 1.

Deep neural networks (DNN) recently advanced the state of the art in multiple machine learning tasks, including computer vision. However, increased performance often goes hand in hand with an increase in model complexity, see for example [7, 31]. Complexity causes a substantial computational effort during training and to a lesser extend at runtime. This trend conflicts with the limited resources available in electronic control units for driver assistance systems and autonomous driving.

In order to improve the overall efficiency, a structure comprised of a shared encoder stage and task specific decoders is introduced. The function of the encoder stage is to generate joint feature maps, that allow simultaneous inference of traffic scene segmentation and recognition. In contrast to two separate task specific models, the integrated architecture avoids redundant computations of feature maps.

To achieve computational efficiency, the feature encoder relies upon the inception-v1 (GoogLeNet) architecture [26] due to its compactness. The decoder for the segmentation task employs a fully convolutional skip architecture [15], whereas the decoder for the task of road topology recognition is comprised of a conventional fully connected classifier. This architecture thus supports simultaneous segmentation and recognition predictions in a single forward pass.

Furthermore, the segmentation decoder is adapted to exploit specific properties present in traffic scene images. More precisely, it exploits spatial priors present in traffic scenes when observed from the perspective of the vehicles onboard camera, as for example the area in front of the car almost always depicts the road, whereas sidewalks and buildings occur mostly on the perimeter of the scene. The Hadamard product in conjunction with element-wise weight matrices efficiently takes advantage of these spatial priors and improves the segmentation performance. The method is evaluated on both tasks w.r.t. the cityscapes benchmark dataset [2]. A subset of the original data is augmented with manually labeled ground truth of road topology.

2 Related Work

Traffic scene recognition has been dominantly studied for traditional computer vision methods, based on feature engineering. [3] analyses a two-step approach, in which a superpixel representation defines characteristic scene features. The classification is carried out w.r.t. distinct road topology classes. [22] examines traffic scene recognition for a vehicle fleet management application. They focus on compact feature descriptors to enable computation in remote back-end systems.

For the general task of image recognition, DNNs recently became state of the art after the seminal paper by [11] in 2012. Other influential publications include [13], which proposes 1×1 convolutions, and [23], which emphasizes the importance of network depth. Building upon these methods, [26] propose the inception-v1 architecture for DNNs, which won the ImageNet Large-Scale Visual Recognition Challenge in 2014 [20]. It is build around submodules dubbed "inception modules". One of its striking features is its efficiency in terms of computational resources. This architecture is later refined in various ways, see [7, 25] introduces the ResNet architecture as a technique to feasibly train very deep DNNs. Herein, residual connections are utilized, to additively merge input and output signals of intermediate layers. The authors provide theoretical foundations, as well as experimental results to justify their approach, presenting a variant of their architecture that consists of 152 layers and 60 million parameters.

Generally, DNNs for image recognition are constructed from stacked processing layers, in which the spatial resolution associated with these layers is successively reduced with increasing depth until a scalar class label is finally predicted. [15] present an end-to-end trainable DNN architecture designed for the task of segmentation, named the "fully convolutional network" (FCN). A recognition DNN is adapted for the task of segmentation, by branching the network at intermediate layers, and combining these branches to form a pixel-dense new output path with preserved spatial resolution. This adaptation approach allows for small size datasets, by employing the technique of "fine-tuning" [29]. Thus, a recognition model provides an initialization for a segmentation model. This is beneficial, as available segmentation datasets are comparatively small, due to the tedious manual segmentation and labeling effort. Building on this, a popular approach consists of combining DNNs and "conditional random field" models (CRFs) to perform semantic segmentation, see [1, 12]. It is generally argued, that DNNs perform especially well for feature representations while CRFs better capture contextual relation modeling.

Recent methods for semantic segmentation rely on the FCN architecture, in conjunction with very deep residual models. [28] refine the ResNet architecture by inserting additional residual units in a parallel manner and dub their approach as wide ResNet. [31] also build upon the ResNet architecture, adding parallel computation of multiple pooling layers of different dimensions. Both approaches achieve state of the art results on the cityscapes benchmark.

An alternative approach is proposed by [8]. The classification and segmentation tasks are decoupled, and independently performed by two DNN models. Both models are trained separately such that a scalar class label is determined first, and the object contours are computed afterwards.[6] introduces an approach, termed "hypercolumn", similar to [15] information from coarsely resolved "deeper" layers and information from finer resolved "early" layers is combined, in order to form pixel descriptors in terms of feature vectors. These descriptors constitute the input for the final classification step to obtain a dense segmentation output.

The joint reasoning about complementary representations with DNNs is examined in [27] for classification problems with small amounts of classes. Environment perception based on a combined recognition and segmentation approach is examined by [19] and employed in [18] for indoor mobile robot control. The approach relies on traditional features, and a superpixel representation is used to solve the segmentation task.

3 Deep Neural Networks

DNN consist of multiple processing layers that typically are connected in a hierarchical structure, such that every layer processes the previous layers output data. The outputs of the nth layer are composed into a three dimensional matrix X_n of dimensions $h \times w \times d$, where h and w define the two dimensional image coordinates and d represents an image channel dimension. For the input data d denotes the color channels and thus $d = 3$. One important aspect of DNN architectures motivated from biology is the concept of receptive fields with limited connections between adjacent layers to a local pixel neighborhood. Figure 2 illustrates the connectivity of receptive fields across multiple layers. The processing layers implement one of several different mathematical operations. Those types of processing layers relevant for the present work are described in the following.

3.1 Processing Layers

Convolutional layer: The normalized weighted sum over all pixels within the input of a receptive field is equivalent to a linear two-dimensional convolution, if the receptive fields share common weight factors. The convolution kernel is denoted as Θ_n. Its reuse across receptive fields leads to a substantial reduction in the number of free parameters, which in turn simplifies the parameter identification w.r.t. memory consumption and required computation time. The outputs of the convolutional layer are defined by the discrete linear convolution according to Eq. 2:

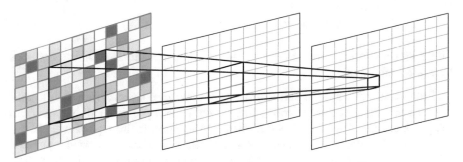

Fig. 2 Concept of receptive fields: connections between adjacent layers are limited to a local neighborhood

$$X_n(h', w', d') = (X_{n-1} * \Theta_n)(h', w', d') \tag{1}$$

$$= \sum_{i=1}^{h} \sum_{j=1}^{w} \sum_{k=1}^{d} X_{n-1}(i, j, k)\Theta(h' - i, w' - j, d' - k) \tag{2}$$

Since input data consists of multiple image channels, the convolution kernel likewise exhibits three dimensional structure. Generally, the filter responses stemming from multiple different convolution kernels are computed, such that the output is comprised of multiple feature map channels. In addition, a transposed convolutional layer is defined by multiplying single pixels with a convolution kernel and combining the resulting values in an output image. In our approach this feature implements an oversampling operation.

RELU-activation layer: In the context of DNN, a separate processing layer is defined for computing non-linear activation functions. For traditional sigmoid functions, the gradient decreases for increasing activations, which is problematic for efficient optimization. For this reason, so called rectified linear units (RELU) are employed as non-linear activation functions:

$$X_n(h', w', d')$$
$$= \max(0, X_{n-1}(h', w', d')) + v \cdot \min(0, X_{n-1}(h', w', d'))$$

RELUS display non-vanishing gradients for the entire range of inputs X_{n-1}. For negative X_{n-1}, the gradient is determined by the parameter v which is set manually. Compared to traditional activation functions, RELU units allow for a more efficient training and also more closely reflect biological processes inspired from neuroscience [5].

Pooling layer: Image pixels are obtained from subsampling within each receptive field, a technique known as "pooling". In our case, subsampling operates with max pooling layers. Let $R_n(h', w', d', X_n)$ be the subset of image pixels within the receptive field centered around (h', w', d'). The pooling layer is formally defined by:

$$X_n(h', w', d') = \max(R_{n-1}(h', w', d', X_{n-1}))$$

Other variants of the pooling layer might be implemented based on averaging or stochastic subsampling.

Fully connected layer: the structure of fully connected layers corresponds to conventional artificial neural networks, since every neuron processes all outputs from the previous layer. Therefore, receptive fields do not apply in this type of layer. Mathematically, the fully connected layer is a multiplication of the vectorized input data with a weight matrix Θ_n, which contains the learnable parameters of this layer. The vectorized matrices are denoted by $\text{vec}(X_n)$, the fully connected layer is defined as:

$$\text{vec}(X_n) = \text{vec}(X_{n-1}) \cdot \Theta_n$$

The number of learnable parameters of the fully connected layer is equivalent to the product of its input and output dimensions. In order to maintain a feasible model complexity, fully connected layers are only applied to significantly sub-sampled image data in deep hierarchical levels.

Hadamard layer: The Hadamard product ∘ performs element-wise multiplication of input feature maps X_{n-1} with a weight matrix Θ_n of equal dimensions. These Hadamard layers can be thought of as 1×1 convolution with position dependent filter kernels. Consequently, they have no effect on the overall receptive field size, as illustrated in Fig. 3. Mathematically, they are given by:

$$X_n = X_{n-1} \circ \Theta_n$$

Hadamard layers encode spatial priors by the means of position dependent weights.

Fig. 3 Connectivity patterns imposed by different layers

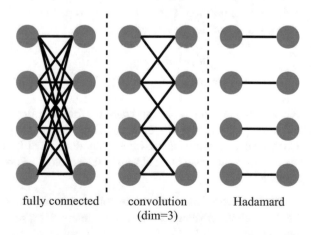

fully connected   convolution   Hadamard
                          (dim=3)

3.2 Parameter Optimization

Mathematically, the training of neural networks constitutes a nonlinear optimization problem. The outputs of the last layer are interpreted as confidence scores for the respective class labels, thus the dimension of image channels d equals the number of categories. The well-known softmax function ρ is applied to the arbitrarily scaled confidence scores to compute class probabilities:

$$\rho(h', w', d') = \frac{e^{X_{n-1}(h',w',d')}}{\sum_{k=1}^{d} e^{X_{n-1}(h',w',k)}}$$

In order to train the network architecture for both, the recognition and the segmentation task, two different training losses are defined for the two separate output paths. Since both outputs depict a classifier, the optimization minimizes the multinomial logistic loss function. In order not to prioritize one task over the other the segmentation loss is normalized with respect to its spatial resolution, see Eq. 4. Thus, both losses produce an error loss in the same range.

The statistics of the cityscapes dataset [2] show, that the class distribution is significantly unbalanced, as few classes account for a large number of examples in the dataset. An easy way to account for unbalanced data, is to perform oversampling of rare classes. However, this is inefficient as in effect it leads to an artificially enlarged training dataset, and an increased computation time per epoch. Instead, each training sample is weighted within the loss function according to its frequency in the dataset. This is straight forward for the recognition loss, in case of the segmentation loss a pixel-dense weight mask enters the loss computation. [21] observe performance deterioration, if class weights are merely adjusted according to their inverse frequencies, and thus propose the "85–15%-rule" introduced in [16] to define class weights. Following [21] the weight γ_c of class c is defined as:

$$\gamma_c = 2^{\lceil \log_{10}\left(\frac{\eta}{f_c}\right) \rceil} \tag{3}$$

where f_c is the frequency of occurence and η is the count of the most frequent classes, that account for at least 85% of the dataset (thus the "85–15%-rule"). With these definitions, the optimization of model parameters Θ for a segmentation-only model with a multinomial logistic loss function is given by:

$$\Theta = \operatorname*{argmin}_{\Theta} \left(-\frac{1}{h \cdot w} \sum_{i=1}^{h} \sum_{j=1}^{w} \gamma_{l(i,j)} \cdot \ln\left(\rho(i,j, \, l(i,j)) \right) \right) \tag{4}$$

where i, j are pixel coordinates, ρ represents the class probabilities computed by the softmax function and l defines the ground truth labels. This loss function is minimized by a high confidence for the ground truth class. For the special case of

$h = w = 1$ the scalar loss for recognition-only models is obtained. For the combined model, the unweighted sum of both loss functions is used.

The training of the DNNs is carried out following the well-known backprogagation algorithm based on the gradient decent on the loss function. In practice, a variant dubbed stochastic gradient descent (SGD) with momentum is employed. This variant differs from the ordinary gradient descent in two aspects. First of all, the output error is computed only on a subset of the overall training dataset. The size of this subset is named batch size and the processing of the complete dataset is dubbed an epoch. Secondly, the gradient decent update rule is extended by a weighted momentum term, such that the weight update is computed according to a combination of the current gradient and previous updates. In effect, this leads to suppression of high frequency fluctuations in the loss function (smoothing). Using the step size η and the momentum α, the update rule for iteration τ is given by:

$$\Delta\Theta_{n,\tau+1} = -\eta\frac{\partial E}{\partial\Theta_n} + \alpha\Delta\Theta_{n,\tau}$$

4 Architecture for Integrated Recognition and Segmentation

Embedded applications, such as driver assistance systems and automated driving, are subject to rigid constraints regarding computation time, energy consumption and memory capacity. It is therefore advantageous to share computations between the separate tasks of recognition and segmentation. For this, the DNN incorporates a joint feature representation, which simultaneously maps those visual cues relevant for the two tasks. The approach is comprised of an encoder-decoder architecture, which employs a shared encoder stage to generate the joint feature maps, and separate decoders compute predictions for their specific tasks. Figure 4 illustrates the architecture.

4.1 Feature Encoder

In the outlined structure, the computational burden is assumed to mostly lie with the computation of the shared feature maps. Therefroe, the feature encoder consists of the two-dimensional layers of the inception-v1 architecture.[1] Without its final classifier stage, this architecture holds less than 6 million parameters. Due to its compactness, it is well suited to enable fast processing in ADAS control units. For the sake of clarity, repeating sub-structures are shown only once, feature map dimensions

---

[1]This work employs a variant of the architecture published at https://github.com/BVLC/caffe/tree/master/models/bvlc_googlenet. Accessed: 18.01.2017.

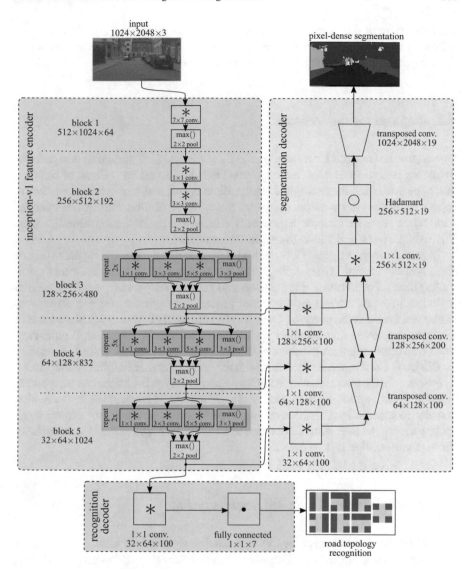

Fig. 4 Scheme of the proposed network architecture. The underlying inception-v1 feature encoder is condensed into subparts of equal spatial resolution. The right branch represents the FCN-based decoder for the *segmentation task*. The bottom branch depicts the fully connected classification decoder for the *recognition task*. The given dimensions represent the respective output feature maps

are given only at the output of intermediate pooling operations and RELU layers following convolution operations are not included in Fig. 4. For further details on the inception-v1 architecture, we refer the reader to [26].

4.2 Segmentation Decoder

Our segmentation decoder rests upon the FCN method [15] particular designed for image segmentation. Herein, a recognition DNN is adapted for the task of segmentation, by branching the network at intermediate layers, and fusing these branches to form a pixel-dense new output path with preserved resolution. Hence, the intermediate layers of the encoder stage generate the feature maps for the segmentation task. The right part of Fig. 4 shows the segmentation decoder.

In contrast to the original FCN method [15], in which the processing paths branch directly after generating the pooling outputs, our network only branches after the following block. Furthermore, [15] proposes to sum the outputs of the added parallel paths to form a pixel-dense segmentation. Instead, our approach employs vector concatenation for the same purpose. Our analysis reveals, that this modification leads to a slightly improved segmentation performance. To improve the overall efficiency, 1×1 convolutions adjust the channel dimensions prior to fusion of feature maps.

Obviously, traffic scenes recorded with the onboard camera exhibit spatial priors, as for example the road and other vehicles appear mostly in the image center, whereas sidewalks and buildings are located at the periphery. Therefore, the segmentation decoder includes a Hadamard layer, with the objective to efficiently exploit those spatial priors. The following section discusses the advantages of the Hadamard layer for the segmentation task in detail.

4.3 Spatial Priors for Scene Segmentation

Typical DNNs employ mappings with limited receptive field sizes. Thus, activations depend on a limited set of inputs within their direct neighborhood. In theory, the stacking of such mappings increases the overall receptive field of the whole model, see Fig. 2. However, typical global receptive field sizes are still far below the image resolution of modern automotive cameras. This holds especially, since receptive fields observed in practice are even smaller than their theoretical sizes. For example, [32] reports global receptive fields with less than 100 pixels in diameter for a typical DNN, similiar observations are reported by [14]. Furthermore, convolutional layers replicate identical filter kernels at every spatial position of an image, due to the weight sharing regularization. This prevents the efficient utilization of spatial priors, as they are neither captured in limited receptive fields nor encoded in position dependent weights.

As outlined before, the utilization of spatial priors is either achieved by increasing the receptive field sizes or by explicitly encoding mean class distributions in position dependent weights. Maximal sized convolution kernels are equal to fully connected layers, their number of parameters is the product of the input and output dimensions. Thus, they cannot be used to compute spatially resolved outputs in practice. [30] proposes dilated filter kernels, which alleviate this problem but instead lead to increased feature map dimensions that are computationally expensive. In addition, the introduced sparse mappings might not be suited to capture class distributions at high spatial resolution. As our objective is to explicitly model mean class distributions, a Hadamard layer is added before the final transposed convolution step in the segmentation decoder stage.

4.4   Recognition Decoder

The recognition decoder predicts the road topology from the feature maps of the encoder stage. Following the original inception-v1 architecture, prediction is based on the final feature map of the encoder stage. After implementing a dimensionality reduction by means of 1×1 convolution, the final road topology predictions are inferred by a fully connected classifier. The recognition decoder is shown in the bottom part of Fig. 4.

5   Experiments

The experimental evaluation is based in the cityscapes dataset [2], which is comprised of urban street scenes and ground truth labels for semantic segmentation. The benchmark includes coarse and fine annotated samples, that are both used for training. In order to analyze the recognition task, 1199 images of the training partition and 400 images of the test partition, corresponding to nine different cities, are manually augmented with road topology ground truth annotations. Since typical data augmentation techniques alter the road topology as well as the mean class distributions in the dataset, no data augmentation techniques are employed. All images are processed at their original resolution. The DNN training is done with the caffe [9] deep learning framework, using the stochastic gradient descent solver with momentum [24]. Training operates with a momentum of 0.9, a fixed learning rate of 5×10^{-4}, a weight decay factor of 5×10^{-4} and a batchsize of 1.

Using the method of fine-tuning [29], all DNN parameters belonging to the original inception-v1 structure are initialized from a model pre-trained on ILSVRC12 [20], and published alongside the works of [9]. During the analysis, the combined model with both outputs is compared against models trained for either only segmentation or only recognition. In the latter case, the obsolete parts of the structure in Fig. 4 are omitted. In addition, models with and without the Hadamard layer are examined in order to analyze its effect on the segmentation performance. Therefore, a total of four models are evaluated hereafter.

During experimentation, it turned out that the best performance for segmentation-only models is achieved after 100000 iterations. Hypothetically caused by the smaller dataset, recognition-only models tend to overfit after about 10000 iterations, thus training is stopped afterwards. The combined task implements the full architecture illustrated in Fig. 4. Best results are obtained by initializing all parameters from the segmentation-only model, and fine-tuning the combined model for 10000 iterations. The combined model is trained w.r.t. the sum of both loss functions.

5.1 Recognition

To examine the recognition task, the recognition-only model and the combined model is analyzed w.r.t. the test partition of the road-topology dataset. The precision, recall and F1-score according to the definition of [4], and averaged over all road topology classes, are reported in Fig. 5.

It is evident, that the recognition-only model generally achieves better results than the combined model. This decrease in performance is explained by the stronger constraints imposed on the feature maps due to the twofold optimization target. Intuitively, a harder problem needs to be solved with a feature encoder of the same capacity. Further comparison with similar works reveals, that both models are able to outperform traditional methods based on feature engineering. For example [3] observes a precision of 0.45 for a similar 8-class recognition task. Additionally, Fig. 5 indicates that the resulting classifier does not overly focus on frequent classes, as no substantial divergence of the error across categories is observed. Presumably, this is attributed to the loss weighting used during parameter optimization.

Furthermore, Table 1 reports the full confusion matrices. The highest accuracy, as well as the highest false-positive rates, are observed for straight streets (a.) and full intersections (g.), which also occur most frequently in the dataset. The biggest decrease in performance for the combined model emerges for curves without branches (b. and c.).

Fig. 5 Average per-class results for the recognition task

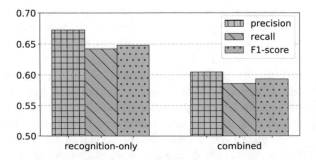

Table 1 Confusion matrices for the recognition task

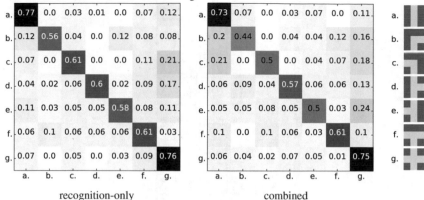

	a.	b.	c.	d.	e.	f.	g.
a.	0.77	0.0	0.03	0.01	0.0	0.07	0.12
b.	0.12	0.56	0.04	0.0	0.12	0.08	0.08
c.	0.07	0.0	0.61	0.0	0.0	0.11	0.21
d.	0.04	0.02	0.06	0.6	0.02	0.09	0.17
e.	0.11	0.03	0.05	0.05	0.58	0.08	0.11
f.	0.06	0.1	0.06	0.06	0.06	0.61	0.03
g.	0.07	0.0	0.05	0.0	0.03	0.09	0.76

recognition-only

	a.	b.	c.	d.	e.	f.	g.
a.	0.73	0.07	0.0	0.03	0.07	0.0	0.11
b.	0.2	0.44	0.0	0.04	0.04	0.12	0.16
c.	0.21	0.0	0.5	0.0	0.04	0.07	0.18
d.	0.06	0.09	0.04	0.57	0.06	0.06	0.13
e.	0.05	0.05	0.08	0.05	0.5	0.03	0.24
f.	0.1	0.0	0.1	0.06	0.03	0.61	0.1
g.	0.06	0.04	0.02	0.07	0.05	0.01	0.75

combined

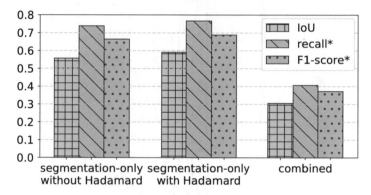

Fig. 6 Average per-class results for the segmentation task, measures marked with * are calculated on the validation dataset

5.2 Segmentation

For the segmentation task the effect of the Hadamard layer is examined, as well as differences between the combined model and segmentation-only models. Following [4, 10], Fig. 6 reports the average IoU-score, the average accuracy and the average F1-score for the three resulting models.

Similar to the recognition task, a decrease in performance for the combined model is evident for the same reasons. Notably, the segmentation-only model with Hadamard layer outperforms the one without Hadamard layer. This observation confirms the significance of spatial priors supporting the segmentation of traffic scenes and demonstrates that FCN networks benefit from element-wise weights.

Additionally, the per-class IoU metric is reported in Table 2. Less frequent classes belonging to objects, that typically cover small areas in traffic scenes, account for the

Table 2 IoU-scores for the segmentation task, I. segmentation-only without Hadamard, II. segmentation-only with Hadamard, III. combined measures marked with * are calculated on the validation dataset

	Road	Sidewalk	Building	Wall	Fence	Pole	Trafficlight	Trafficsign	Vegetation	Terrain	Sky	Person	Rider	Car	Truck	Bus	Train	Motorcycle	Bicycle	Per Pixel	Per Class	# Parameters(M)	Runtime[a] (s)
I.	0.95	0.65	0.86	0.32	0.36	0.43	0.45	0.54	0.88	0.64	0.92	0.52	0.37	0.87	0.28	0.42	0.35	0.23	0.55	0.80*	0.56	6.2	0.13
II.	0.96	0.70	0.86	0.32	0.37	0.39	0.47	0.53	0.87	0.65	0.92	0.66	0.42	0.89	0.33	0.49	0.41	0.42	0.56	0.80*	0.59	8.7	0.13
III.	0.86	0.40	0.70	0.02	0.07	0.25	0.13	0.31	0.81	0.45	0.83	0.27	0.00	0.46	0.00	0.05	0.02	0.00	0.16	0.71*	0.31	10.2	0.16
[15]	0.97	0.78	0.89	0.35	0.44	0.47	0.60	0.65	0.91	0.69	0.94	0.77	0.51	0.93	0.35	0.49	0.47	0.52	0.67	n/a	0.65	>100	0.5
[17]	0.97	0.78	0.88	0.47	0.44	0.29	0.44	0.55	0.89	0.67	0.93	0.71	0.49	0.91	0.56	0.67	0.57	0.48	0.58	n/a	0.65	>100	4
[30]	0.98	0.79	0.90	0.37	0.48	0.53	0.59	0.65	0.92	0.69	0.94	0.79	0.55	0.93	0.45	0.53	0.48	0.52	0.66	n/a	0.67	>100	4

[a]For this work, runtime was measured on a GTX Titan X using [9].

Fig. 7 Exemplary segmentations, I. input with ground truth, II. segmentation-only without Hadamard, III. segmentation-only with Hadamard, IV. combined

majority of the decrease in performance. This is consistent with the relatively small difference in the per-pixel average of the IoU metric. Figure 7 shows exemplary segmentation results from the validation data for all three models. Again, the subjectively noisier segmentation of the combined model is notable, however dominant classes such as the road are still represented fairly accurate.

In general it is noticeable, that larger models further improve the segmentation performance. However when comparing the average runtime as well as the overall number of parameters in Table 2, it is apparent that the present work achieves significantly faster computation times and is thus better suited to enable fast processing in ADAS. Furthermore, the relatively small increase in runtime for the combined model proves the advantage of the shared feature encoder design, enabling predictions on complementary tasks at low additional costs.

Figure 8 shows a heatmap illustration of the learned Hadamard weights from the segmentation-only model, for exemplary classes. Distinctive spatial priors are evident for the road area, as well as for sidewalks. As intuitively expected both cover opposite areas of traffic scenes. The results are less conclusive for cars, strong spatial priors on the outer image parts presumably stem from parking spaces. Trains or trams in urban environments mostly occur in the left area. Since only right-hand traffic is captured in the dataset, this is supposedly due to trams driving in the center lanes in most cities.

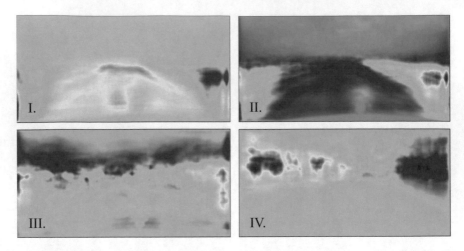

Fig. 8 Heatmap visualization of the Hadamard weights for exemplary classes, I. road, II. sidewalk, III. car, IV. train

6 Summary and Outlook

This work presents an approach for traffic scene recognition in terms of the road topology and scene segmentation with respect to common traffic objects. An integrated DNN architecture with two separate output paths is proposed. The model simultaneously predicts the road topology as well as a pixel-dense image segmentation. The final architecture accounts for limited computational resources in automotive applications.

While strong constraints imposed on the DNN feature maps due to the twofold loss function constitute a complex optimization problem, experimental results on the cityscapes dataset demonstrate, that a meaningful representation of traffic scenes is obtained. This applies in particular to applications, that rely mostly on the segmentation of large traffic objects such as the road area.

Future work is concerned with the comparative analysis of the presented method with very deep architectures such as [7], in terms of both performance and computational effort.

Acknowledgements The funding for this work was provided by the European Regional Development Fund (ERDF).

References

1. Chen LC, Papandreou G, Kokkinos I, Murphy K, Yuille AL (2015) Semantic image segmentation with deep convolutional nets and fully connected crfs. In: 3rd international conference on learning representations. arXiv:1412.7062

2. Cordts M, Omran M, Ramos S, Rehfeld T, Enzweiler M, Benenson R, Franke U, Roth S, Schiele B (2016) The cityscapes dataset for semantic urban scene understanding. In: Proceedings of the IEEE conference on computer vision and pattern recognition, pp 3213–3223

3. Ess A, Müller T, Grabner H, van Gool L (2009) Segmentation-based urban traffic scene understanding. In: Proceedings of the 20th British machine vision conference, pp 84–1

4. Fritsch J, Kühnl T, Geiger A (2013) A new performance measure and evaluation benchmark for road detection algorithms. In: Proceedings of the 16th IEEE conference on intelligent transportation systems, pp 1693–1700

5. Glorot X, Bordes A, Bengio Y (2011) Deep sparse rectifier neural networks. In: Aistats, vol 15, p 275

6. Hariharan B, Arbeláez P, Girshick R, Malik J (2015) Hypercolumns for object segmentation and fine-grained localization. In: Proceedings of the IEEE conference on computer vision and pattern recognition, pp 447–456

7. He K, Zhang X, Ren S, Sun J (2016) Deep residual learning for image recognition. In: Proceedings of the IEEE conference on computer vision and pattern recognition, pp 770–778

8. Hong S, Noh H, Han B (2015) Decoupled deep neural network for semi-supervised semantic segmentation. In: Advances in neural information processing systems, vol 28. MIT Press, pp 1495–1503

9. Jia Y, Shelhamer E, Donahue J, Karayev S, Long J, Girshick R, Guadarrama S, Darrell T (2014) Caffe: convolutional architecture for fast feature embedding. In: Proceedings of the 22nd ACM international conference on multimedia. ACM, pp. 675–678

10. Kendall A, Badrinarayanan V, Cipolla R (2015) Bayesian segnet: model uncertainty in deep convolutional encoder-decoder architectures for scene understanding. arXiv:1511.02680

11. Krizhevsky A, Sutskever I, Hinton G (2012) Imagenet classification with deep convolutional neural networks. In: Advances in neural information processing systems, pp 1097–1105

12. Lin G, Shen C, van den Hengel A, Reid I (2016) Efficient piecewise training of deep structured models for semantic segmentation. In: Proceedings of the IEEE conference on computer vision and pattern recognition. IEEE, pp 3194–3203

13. Lin M, Chen Q, Yan S (2013) Network in network. arXiv preprint. arXiv:1312.4400

14. Liu B, He X, Gould S (2015) Multi-class semantic video segmentation with exemplar-based object reasoning. In: Proceedings of the IEEE winter conference on applications of computer vision, pp 1014–1021

15. Long J, Shelhamer E, Darrell T (2015) Fully convolutional networks for semantic segmentation. In: Proceedings of the IEEE conference on computer vision and pattern recognition, pp 3431–3440

16. Mostajabi M, Yadollahpour P, Shakhnarovich G (2015) Feedforward semantic segmentation with zoom-out features. In: Proceedings of the IEEE conference on computer vision and pattern recognition, pp 3376–3385

17. Papandreou G, Chen LC, Murphy K, Yuille AL (2015) Weakly-and semi-supervised learning of a dcnn for semantic image segmentation. In: Proceedings of the IEEE international conference on computer vision, pp 648–656

18. Posada LF, Hoffmann F, Bertram T (2014) Visual semantic robot navigation in indoor environments. In: Proceedings of the 41st international symposium on robotics, VDE, pp 1–7

19. Posada LF, Narayanan KK, Hoffmann F, Bertram T (2013) Semantic classification of scenes and places with omnidirectional vision. In: Proceedings of the IEEE European conference on mobile robots, pp 113–118

20. Russakovsky O, Deng J, Su H, Krause J, Satheesh S, Ma S, Huang Z, Karpathy A, Khosla A, Bernstein M, Berg AC, Fei-Fei L (2015) ImageNet large scale visual recognition challenge. Int J Comput Vis 115(3):211–252. doi:10.1007/s11263-015-0816-y

21. Shuai B, Zuo Z, Wang B, Wang G (2016) Dag-recurrent neural networks for scene labeling. In: Proceedings of the IEEE conference on computer vision and pattern recognition, pp 3620–3629

22. Sikirić I, Brkić K, Krapac J, Šegvić S (2014) Image representations on a budget: traffic scene classification in a restricted bandwidth scenario. In: Proceedings of the IEEE intelligent vehicles symposium, pp 845–852

23. Simonyan K, Zisserman A (2014) Very deep convolutional networks for large-scale image recognition. arXiv:1409.1556
24. Sutskever I, Martens J, Dahl G, Hinton G (2013) On the importance of initialization and momentum in deep learning. In: Proceedings of the 30th international conference on machine learning, pp 1139–1147
25. Szegedy C, Ioffe S, Vanhoucke V, Alemi A (2016) Inception-v4, inception-resnet and the impact of residual connections on learning. arXiv:1602.07261
26. Szegedy C, Liu W, Jia Y, Sermanet P, Reed S, Anguelov D, Erhan D, Vanhoucke V, Rabinovich A (2015) Going deeper with convolutions. In: Proceedings of the IEEE conference on computer vision and pattern recognition, pp 1–9
27. Teichmann M, Weber M, Zoellner M, Cipolla R, Urtasun R (2016) Multinet: Real-time joint semantic reasoning for autonomous driving. arXiv:1612.07695
28. Wu Z, Shen C, Hengel Avd (2016) Wider or deeper: revisiting the resnet model for visual recognition. arXiv:1611.10080
29. Yosinski J, Clune J, Bengio Y, Lipson H (2014) How transferable are features in deep neural networks? In: Advances in neural information processing systems, vol 27. MIT Press, pp 3320–3328
30. Yu F, Koltun V (2016) Multi-scale context aggregation by dilated convolutions. In: 4th International conference on learning representations. arXiv:1511.07122
31. Zhao H, Shi J, Qi X, Wang X, Jia J (2016) Pyramid scene parsing network. arXiv:1612.01105
32. Zhou B, Khosla A, Lapedriza A, Oliva A, Torralba A (2015) Object detectors emerge in deep scene cnns. In: 3rd International conference on learning representations. arXiv:1412.6856

Fuzzy Ontology Support for Knowledge Mobilisation

Christer Carlsson

Abstract Classical management science is making the transition to analytics, which has the same agenda to support managerial planning, problem solving and decision making in industrial and business contexts but is combining the classical models and algorithms with modern, advanced technology for handling data, information and knowledge. We run a knowledge mobilisation project as a joint effort by Institute for Advanced Management Systems Research, and VTT Technical Research Centre of Finland. The goal was to mobilise knowledge stored in heterogeneous databases for users with various backgrounds, geographical locations and situations. The working hypothesis of the project was that fuzzy mathematics combined with domain-specific data models, in other words, fuzzy ontologies, would help manage the uncertainty in finding information that matches the users' needs. In this paper, we describe an industrial application of fuzzy ontologies in information retrieval for a paper machine where problem-solving reports are annotated with keywords and then stored in a database for later use. One of the key insights turned out to be that using the Bellmann-Zadeh principles for fuzzy decision-making are useful for identifying keyword dependencies in a keyword taxonomic tree.

1 Introduction

In this paper, we will deal with a classical problem in industrial management—to effectively solve problems in industrial processes by using experience and documented insight on successful problem-solving procedures. The insight comes from experts and experienced engineers who have found good ways to deal with often complex (and costly) problems, the documents they have written are quite often cryptic and hard to follow for outsiders or less experienced process operators. Then there is the further challenge, in the era of "big data" the documents reside in

C. Carlsson (✉)
Institute for Advanced Management Systems Research, Turku, Finland
e-mail: christer.carlsson@abo.fi

© Springer International Publishing AG 2018
S. Mostaghim et al. (eds.), *Frontiers in Computational Intelligence*, Studies in Computational Intelligence 739, https://doi.org/10.1007/978-3-319-67789-7_8

databases of 10,000 documents, sometimes hundreds of thousand documents, which makes it challenging to retrieve them, especially when problems are real and urgent. If and when the documents are retrieved the accuracy and effectiveness of the problem-solving processes reported needs to be verified and validated, i.e. the problem solvers should not be guided to follow processes that are inappropriate or incomplete for the tasks at hand.

In industrial management, we have a tradition of following management science methodology—we aim to find the best or optimal solutions. When faced with problems of the kind we have sketched above (something is going (very) wrong in an industrial process) the process is straightforward (cf. [8]):

(i)   work out a relevant and focused description and definition of the problem at hand;
(ii) build mathematical models to capture the essence of the problem;
(iii) work out algorithms that can be used as problem solvers in the framework of the models;
(iv) simulate and try alternative solutions with the algorithms;
(v)   find an optimal solution;
(vi) implement the optimal solution for the original problem and find out if it actually works; if not, go back to (i) and restart the process.

This process has served us very well for more than six decades and it has generated thousands of models and algorithms that have helped work out solutions to complex and difficult problems in industry. Digitalisation, which brought the "big data" and "fast data (streaming big data)" challenges, combined with advances in information systems technology has initiated discussions about the relevance of management science and its methodology. The use of mathematical models and algorithms appears to be too burdensome for modern managers—they believe in on-line and real time information support for fast decision-making. We have worked out some principles for *knowledge mobilisation* (cf. [11]) that we propose to be useful as a combination of the good and useful principles of management science with the fast knowledge retrieval that modern managers expect.

This paper is structured as follows: in Sects. 1.1 and 1.2 a summary sketch outlines management science and analytics; in Sect. 2 we work out fuzzy ontology for industrial applications; in Sect. 3 we show how digital coaching could assist users of fuzzy ontology; Sect. 4 is a summary and conclusions.

1.1 Management Science

Management science (MS) methodology—and especially operations research (OR) that applied the same methodology for engineering problems and theory development—was first attacked in the early 1970s (cf. [8]) for failing to deal with the real world problems managers have to tackle, for oversimplifying problems and for spending too much time with mathematically interesting but practically irrelevant problems and solutions. The message was simply that management science methodology produced theory and methods that were irrelevant for handling actual management problems. In a paper in 1984 (cf. [8]) I argued that fuzzy sets—when properly worked into management science methodology—would make the models, the algorithms and the theory more relevant and better suited to deal with management problems in practice.

Now, more than thirty years later, I have to admit that we were not successful in bringing it about, that fuzzy sets remained a marginal development in management science. Management scientists working with fuzzy sets theory and methods managed to get fuzzy sets based methods accepted only for some limited applications, such as multiple criteria optimisation, real options valuation, logistics optimisation, etc. The main reasons for acceptance were the algorithmic benefits of allowing the use of fuzzy numbers, the logic and core of fuzzy sets has not been widely accepted in management literature.

Management science and operations research have also changed over the decades. Two major organisations in the field—TIMS and ORSA—merged and became INFORMS to combine the applications oriented research (TIMS) with the algorithms and theory oriented research (ORSA). The 2010s now sees a renaissance of the MS/OR methods—the annual INFORMS conferences collect 2-3000 participants; in Europe the EURO Association (a sister organisation) also has 2-3000 participants at the annual EURO conferences. Both organisations run major, well-established journals with high impact factors and there are dozens of journals publishing material produced under guidance of management science methodology. The field is alive and well and promotes lively research that activates thousands of researchers. In order to achieve this the field only had to change name—*analytics*—and to be promoted by Harvard Business School scholars and publications (cf. [17]). Of course, more than the name changed, also the settings in information systems and technology, the digital context for modern business, the "big data" challenges that created an urgent need for mathematical and statistical tools, and the skills and knowledge level of the users are key reasons for the surge in analytics modelling and implementations. There is a role for fuzzy sets theory and methods in analytics even if most of the people who work with and promote analytics do not know much (if anything) about the possibilities with fuzzy algebra and algorithms based on that algebra.

1.2 Analytics

Operations Research and Management Science are now in the process of transformation by (Business) *Analytics*, which is getting the attention of major corporations and senior management. On our part, in our work with complex, difficult problems for large industrial corporations, we have for a number of years been promoting *Soft Computing* to the same audience instead of trying to explain fuzzy sets theory and fuzzy logic in the way it was originally done. The experience we have—summarized in a few words—is that *analytics* methods, which implement *soft computing* theory and algorithms are turning out to be very effective and useful for planning, problem solving and decision making in "big data" environments (cf. [15]). *Analytics* adds value to management; it promotes data-driven and analytical decision making, which was somehow "reinvented" as being important and useful for management that had relied on other schools of thought for a couple of decades. Analytics builds on recent software improvements in information systems that has made data, information and knowledge available in real time in ways that were not possible for managers only a few years ago (Davenport and Harris [17]). INFORMS gradually found out that the new movement represents both "potential opportunities" and "challenges" to management science and operations research professionals (Liberatore and Luo [27]). The methods and the application cases worked out in the Davenport-Harris book are very close to traditional management science methodology, actually so close that a manager probably fails to see any differences, which is why INFORMS finds "challenges".

Soft Computing (introduced by Lotfi Zadeh in 1991) builds on fuzzy sets theory [34], fuzzy logic, optimisation, neural nets, evolutionary algorithms, macro heuristics and approximate reasoning. Soft Computing is focused on the design of intelligent systems to process uncertain, imprecise and incomplete information; soft computing methods applied to real-world problems offer more robust, tractable and less costly solutions than those obtained by more conventional mathematical techniques.

Liberatore and Luo [27] list four factors that drive the analytics movement: (i) availability of data, (ii) improved analytical software, (iii) the adoption of a process orientation by organisations, and (iv) managers and executives who are skilled users of information and communication technology. If we compare these factors to the experience of the 1980s the last factor is probably the most important driver—there is a new generation of managers and executives in charge of the corporations that are using information technology as part of their daily routines. They work with data, information and knowledge on a real time basis and they continuously hunt for improved analytical tools to help give them competitive advantages. They do not necessarily recognize the analytical tools as classical management science algorithms. Analytical software (cf. (ii)) has become user-friendly with graphical user interfaces and visualisation of results; users typically do not have the mathematical background to get into details with the algorithms. Information technology has made data available on a real time basis, which allows online planning, problem solving and decision-making. Maybe "allow" is not the right verb as online management work in real time is more of a necessity to keep up with the competition. The same driver also explains the adoption of a process orientation (cf. (iii)) as management work typically is group- and teamwork online and in real time. Davenport and Harris [17] describe analytics as "the extensive use of data, statistical and quantitative analysis, explanatory and predictive models and fact-based management to drive decisions and actions". Liberatore and Luo [27] identify three levels of modeling—descriptive, predictive and prescriptive—and state that Management Science and Operational Research typically would focus on advanced analytics, i.e. prescriptive modelling. They also point out that analytics would focus on the transformation of data into actions through analysis and insight, which in their discussion contributes to the application cases of management science.

The modern movement of *analytics* appears to offer interesting possibilities and opportunities for soft computing. The movement is data-driven, which will require tools for handling *imprecision*; the movement focuses on managers, who need to deal with real world problems, for which available data, information and knowledge are incomplete, imprecise and uncertain but should allow for fast, often intuitive conclusions. The movement builds on improved analytical software that offers platforms for a multitude of algorithms, intelligent technologies, soft computing, computational intelligence, etc. Modern analytics offers platforms and environments for *digital coaching* of managers in planning, problem solving and decision-making.

There are benefits of having worked with management science for a few decades (like the present author and my contemporaries)—there has been hundreds of innovative ideas and some successful solutions, from which it has been possible to extract working principles and a growing understanding of how good science can guide and contribute to successful planning, problem solving and decision-making. In the context of the digital economy these processes—not surprisingly—offer new challenges: real-time management challenged by "big data" and relying on fast processing by advanced analytics methods would best be carried out by postdoc-qualified managers—these are rather scarce and would most often be filtered out by corporate career qualifying processes much before they reach senior management positions. Thus there will be a need to reinstate "coaching" functions with the advanced analytics methods to tell/explain to the users what can/should be done, how it should be carried out, what the results are and what they mean, and how they should be applied (with explanations of what could/should not be done).

I have worked out my storyline in the context of analytics and soft computing and the history that has formed that context over the last few decades. I will work with fuzzy ontology modelling, one of the more advanced analytics methods with a theory that managerial decision makers cannot easily apply, but which shows results that will be useful for the knowledge mobilisation we have in mind.

2 Fuzzy Ontology in Industrial Applications

Fuzzy ontology offers solutions for addressing semantic meaning in an uncertain and inconsistent world. Since many concepts that are needed for intelligent systems lack well defined boundaries, or precisely defined criteria of membership, we need fuzzy logic to deal with notions of vagueness and imprecision. As with fuzzy logic, reasoning is approximate rather than precise. The aim is to avoid the theoretic pitfalls of monolithic ontologies, to facilitate interoperability between different and independent ontologies [16], and to provide flexible information retrieval capabilities [1, 2, 25, 26]. We have developed concepts for a tool that will be useful for searching for paper machine knowledge with a search engine based on a fuzzy ontology. The usage scenario for the tool is that a process expert or an experienced engineer, who is handling a problem in the process chemistry of a paper machine, wishes to find past problem solving cases of a similar setting in order to find possible solutions to a current issue. This setting is universal: pieces of knowledge, called "nuggets", are written and stored by companies on different domains in the form of incident reports.

In the industrial project it turned out that the paper machine represents a large domain (cf. Fig. 2) and we decided to select one part of the domain in order to work out some good principles and then to find ways to reuse the principles and scale up to other parts of the paper machine. For the application worked out here we have focused on the chemistry of the "wet end" in order to limit the work effort needed for the domain ontology and to concentrate on a subject on which domain expertise and actual data were available. "Nuggets" are documents that can contain all kinds of raw data or multimedia output extracted from different information systems. An expert author annotates the nuggets with suitable keywords, and then we base the search for documents on these keywords. In addition to providing exact results to queries, the tool uses a fuzzy domain ontology to extend the query to related keywords (cf. Fig. 1). As a result, the search results include nuggets that may not necessarily deal with exactly the same process equipment, variable, function or chemical, but nuggets that may still provide valuable insight to solving the problem at hand. The terms defined in the ontology should be familiar to the users of the system, e.g. to plant operators, process developers and chemical suppliers. The concepts and terminology that experts develop may not be understandable to outsiders, which became an issue with novice users; another problem is the imprecision and shortcuts that experts thrive on as "real experts can figure it out". Nevertheless,

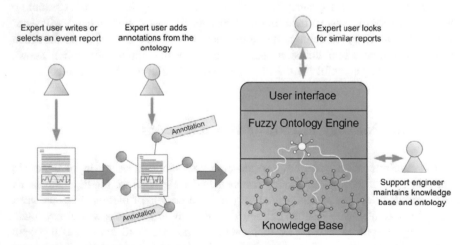

Fig. 1 System concept—a knowledge base of event reports

during the process of building the ontology we found out that the terminology should build on sound principles of conceptual modelling and ontology development.

Furthermore, the goal should be to have common concepts and plant models with equipment manufacturers and engineering contractors for paper machines that provide the initial information and often participate in modifications and upgrade projects. This creates a link to the ongoing development of engineering data models for various industrial areas. Fuzzy keyword ontologies should thus make use of relevant product and plant modelling standards and more general, higher level ontologies. The case with the paper machine contributes to a more general research objective: *to find out if fuzzy ontology can contribute to and enhance knowledge retrieval in the process industry domain.*

2.1 Fuzzy Ontology for Process Industry

Problems with and around the paper machine generate *event reports* that describe what happened, how the problem was tackled, what materials, components, functions etc. were involved, and what solutions were tried and tested, what solutions worked and what the results were. The events (and event reports) are identified with keywords and the task is to build an ontology of keywords that can identify (and help locate) both the relevant event report and similar reports. The ontology construct worked with both engineering and operational knowledge of a paper machine and used the following taxonomic system,

- Top layer: general concepts (based on international standards).
- Middle layer: vocabulary defined and shared by business partners to share knowledge of, (e.g.), the type and structure of process equipment. This layer extends the top layer with domain-specific keywords.
- Bottom layer: custom, company-specific concepts, e.g. specific products and component types, or even individual process plants.

The system of keywords represents concepts, properties, relationships, axioms, and reasoning schemes relevant for the application area. Based on various upper ontologies and industrial data models we identified the following keyword categories as needed to characterize event reports:

- Systems: types of real-world components of a process plant, e.g. machines, buildings, software and people (cf. Fig. 2).
- Functions: phenomena and activities carried out at an industrial plant.
- Variables: properties and state variables of various entities, e.g. temperature.
- Events: types of interesting periods of plant life described in event reports e.g. test runs or equipment failures.
- Materials: raw materials, products, consumables, etc. handled in a process plant.

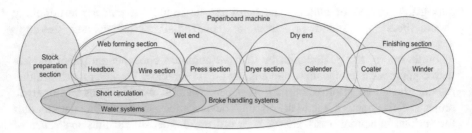

Fig. 2 System example—a fuzzy decomposition of a paper making line

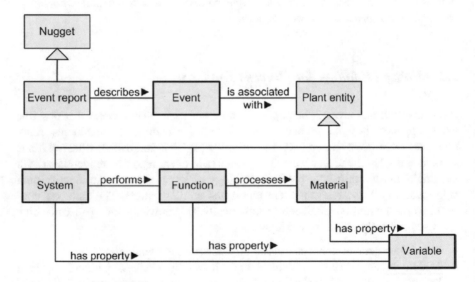

Fig. 3 Overall domain concepts

The basic approach to conceptualize the application is an informal Unified Modelling Language (UML) class diagram above (cf. Fig. 3).

Event reports describe events that are related to various entities of the paper machine, e.g. to equipment, processing functions and materials. There are no assumptions about the internal structure of event reports; an expert characterizes them with keywords selected from a fuzzy ontology. The expert can select the keywords from five categories: event, system, function, material and variable. All keywords represent an entity type and can have subtypes and smaller parts; keywords are used to characterize event reports and other nuggets stored in a database.

Classification (*is-a*) and decomposition (*part-of*) can be found in most ontologies and data models. They are important in the industrial context as well. The keywords in each category are linked by *is-a* and *part-of* relationships as illustrated in Fig. 4.

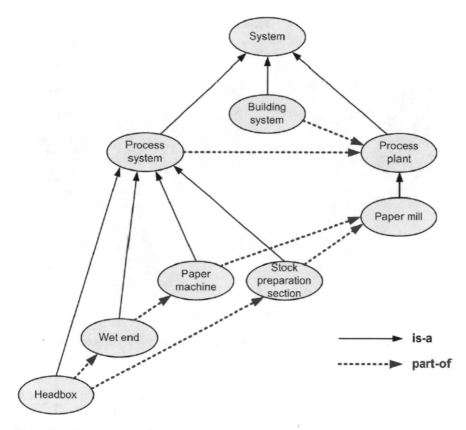

Fig. 4 Simplified event classification matrix

Furthermore, the ontology should model functional and other kinds of dependencies between keywords in various keyword categories. As an example, *systems* work for some purposes, i.e. they play various roles in carrying out one or more functions. This creates a link between the keywords "*wire section* (a part of a paper machine)" and "*formation* (a quality measure of the produced paper)". Modelling classifications, decompositions and various dependencies leads to a situation where we have a taxonomy tree for each keyword category and a set of part-of relationships trees describing the decomposition to various domain entities (Fig. 5). In addition, there are dependency relationships linking keywords to each other.

2.2   The Keyword Ontology

In order to develop a software tool the ontology should be expressed and stored in a more formal way. The basic approach for representing a fuzzy ontology is a

Taxonomy trees

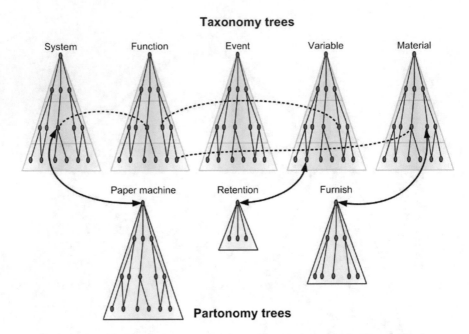

Partonomy trees

Fig. 5 The fuzzy ontology defines classification, decomposition and miscellaneous dependency relationships between keywords

combination of UML class and instance diagrams (cf. Fig. 6). From all kinds of nuggets, the demonstration only looks at event reports that usually describe problematic situations. Each report, like the specific instance "*Report #1*" in the diagram, is related to types of systems, plant functions, events, etc. Only the reference "*Report #1*" to the keyword "*Holes*" is shown in Fig. 6. As indicated by the "*instance of*" associations, all keywords seen by the users are individuals, i.e. instances of a subclass of "Keyword category". Fuzzy dependencies between keyword instances are described by a few fundamental relationship types like *is-a* (specialization), *part-of* and, as an option, *instantiation*. In the first version we focus on "*specializations*", i.e. fuzzy classifications of keywords. The degree of overlapping (or inclusion) of the sets represented by the keywords is described by linguistic labels like "*moderate*" or "significant". The instance named "*Specialization #1*" in Fig. 6 tells us that "*Holes*" is "*to a large extent*" understood as a subclass of "*Quality problem*" but only represents a minor part of its scope. In addition, the keyword "*Holes*" may also specialize other problem types. Here we have modelled dependencies with one single dependency relationship; for simplicity it is symmetric and has a strength value between zero (independence) and one (full dependence).

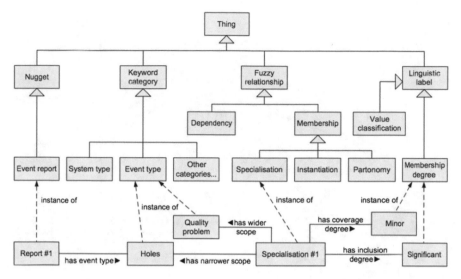

Fig. 6 Representing fuzzy keyword ontology with object classes and their instances

2.3 A Minimax Approach to Assess Keyword Dependencies

We worked out a method for approximating keyword dependencies in the keyword ontology. This method uses Bellman-Zadeh's principles for fuzzy decision-making [3]. Figure 7 shows an event type fuzzy taxonomy. For example, consider the second column of the event classification matrix "*Problem*". All "*Technical problems*" are "*Problems*" and they represent around 80% of all possible problems. That is, "*Technical problem*" covers "*Problem*" with the grade 0.8. Similarly, "*Human errors*" are "*Problems*" and they represent around 30% of all possible problems. That is, "*Human error*" covers "*Problem*" with the grade 0.3. Following Bordogna and Pasi [5] we will assume that if A and B are two keywords in the keyword taxonomic tree, then

$$\text{coverage}(A, B) = \text{fuzzy inclusion}(A, B) \tag{1}$$

For example, "*System fault* = {*Device fault, Design flaw*}" that is, "*System fault*" is a union of these two events. Furthermore, "*Function failure* = {*Design flaw, Drift, Oscillation*}" that is, "*Function failure*" is the union of these three events. "*Design flaw*" covers "*System fault*" with the grade 0.6 and at the same time "*Design flaw*" covers "*Function failure*" with the grade 0.4 (cf. Fig. 7). Moreover, "*System fault*" and "*Function failure*" do not have any more component in common. We compute the degree of dependency between"*System fault*" (*SF*) and "*Function failure*" (*FF*) as their joint coverage by "*Design flaw*" (*DF*)

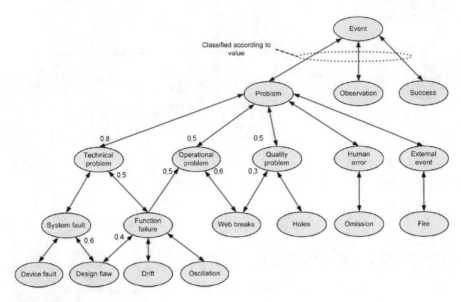

Fig. 7 A fragment of the event type fuzzy taxonomy

$$\text{dependency}(SF, FF) = \min\{\text{coverage}(DF, SF),\ \text{coverage}(DF, FF)\},$$

that is,

$$\text{dependency}(SF, FF) = \min\{0.6,\ 0.4\}\} = 0.4 \tag{2}$$

It is easy to see in Fig. 7 that keywords *"Function failure"* and *"Fire"* are independent since they do not have any component in common. In this case, we have,

$$\text{dependency}(``Function\ failure",\ ``Fire") = 0 \tag{3}$$

Zero means independence, one means full dependence, and values between zero and one denote intermediate degrees of dependency between keywords. It can happen that two keywords have more than one joint component. Then we apply the Bellman-Zadeh principle (a max-min approach) to measure their dependency. For example, suppose that *"System fault"* and *"Function failure"* were to have two joint components, where the first one is *"Design flaw"* and the second one *"Fluctuation"* that has coverage values 0.7 and 0.5, respectively. Then we measure the grade of dependency between *"System fault"* and *"Function failure"* as,

$$\text{dependency}\{SF, FF\} = \max\{\min\{0.6,\ 0.4\},\ \min\{0.7,\ 0.5\}\} = \max\{0.4,\ 0:5\} = 0.5$$

Assume that experts have given all the coverage degrees. Then we can summarize our algorithm as follows: *Compute the degrees of dependency between keywords on the immediate upper level using the max-min approach. Then repeat this procedure until the top layer.*

For example, consider keywords *"Technical problem"* (*TP*) and *"Operational problem"* (*OP*).

Then we find (cf. Fig. 7),

$$\text{dependency}\{TP, OP\} = \min\{\text{coverage}(FF, TP), \text{coverage}(FF, OP)\} = 0.5$$

One can further improve this model by introducing degrees of inclusion and coverage between concepts as suggested by Holi and Hyvönen [21], and Holi [22].

2.4 Demo Architecture and Implementation

The VTT Technical Research Centre of Finland implemented the component-based architecture using the Protégé ontology editor to maintain the fuzzy ontology in OWL format (cf. Fig. 8). The GUI (Graphical User Interface) component guides the user in specifying the information query, and presents the results. There are tools for browsing and evaluating the fuzzy reasoner component directly. A database adapter is used to access report data, which in this case was stored locally in XML files. Similarly, an ontology adapter provides access to the fuzzy ontology, in this case stored in OWL files.

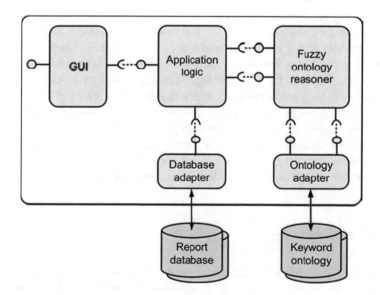

Fig. 8 Component-based demo application architecture [33]

The adapters help hide the different interfaces and protocols of different data sources (e.g. SQL, HTTP) and provide transparent access via an agreed interface. The fuzzy ontology reasoner component is used to process ontology-based information. Its main function is to extend a list of query keywords to a list of their closest neighbours in terms of fuzzy ontology relationships. The component interface also provides methods for directly accessing the ontology concepts and relationships for maintenance and evaluation purposes.

Finally, the application logic component binds all the functionality together by taking the query, using the reasoner component to extend it, passing the extended query to the report database and then combining and ordering the results for the GUI. The fuzzy ontology with fuzzy concepts, relations, and instances was defined using Protégé. The developed ontology moved to the reasoning software as a standard OWL file. Protégé is a widely used tool for developing ontologies and offers some advantages such as a forms-based interface for editing the basic classes and adding the individual keywords, reports, as well as their relationships. Protégé does not have a built-in support for modelling uncertainties, but Straccia [32] introduced a new plug-in called SoftFacts, which is an ontology mediated top-k information retrieval system over relational databases.

Then, in summary, we have examined the possible use of fuzzy ontologies in the retrieval of stored reports that contain knowledge about a paper machine. Since no well-established definitions or mature tools exist, we adopted a practical approach —defining those constructs that we deemed necessary for adequately capturing domain semantics, and implementing an application to test the constructs against actual industrial data. We tested and evaluated the application with the help of industry experts. Although the extended query tool seems to be promising, there has been a great deal of worry about the amount of work needed for maintaining the fuzzy ontology relationship weights.

The builders of models, algorithms and ontologies normally find the functionality and the use of the constructs intuitively clear and self-evident. We found out that this is not the case for the actual users of the knowledge mobilisation tools. Nothing is self-evident, the GUI (cf. Fig. 8) is relatively easy to explain and notes can be added to it to guide the user; after a few iterations with the GUI it becomes routine. The fuzzy keyword ontology is a different matter, like the application logic and the fuzzy ontology reasoner. The users normally want to experiment with and get an intuitive understanding of the tools they are using; in the present case, they need guidance from the system itself. It appears that *digital coaching* is a possible answer and we will explore its possibilities in the next section.

3   Digital Coaching

We tested the knowledge mobilisation potential with the fuzzy keyword ontology using the tools described in Fig. 8. It soon turned out that we had expected too much from the operators in the paper machine monitoring and control room—our perceptions of how the search for and retrieval of event reports (and reports similar to them) could be carried out did not match with the real world handling of incidents and problems. It is now apparent that we should have introduced support tools with the system, what we now call *digital coaches*. The kind of support needed is different for different parts of the system (cf. Fig. 8):

- the GUI needs operating instructions and explanations of functions: {Coach$_i$};
- the Application logic needs advice on similarities of keywords: {Coach$_j$};
- the Fuzzy ontology reasoner needs support on composition of search logic: {Coach$_k$}.

The digital coaching systems got started a few years ago as an answer to the demand on human operators to master advanced automated systems needed to monitor and control often complex and very large industrial process systems. Digital coaching will work on and with data, information and knowledge that is collected from digital devices, instruments, tools, monitoring systems, sensor systems, software systems, data and knowledge bases, data warehouses, etc. and then processed to be usable for the digital systems that will guide and support users. Digital fusion is key to the processing and operates in three phases (cf. [11, 13–15]: data, information and knowledge fusion.

Data fusion is the first step; a function that combines multiple tuples is called a fusion function and the standard, rather simple operation is a fusion of data attributes. The traditional way is to define some ordering relation a priori and then to keep it updated for continuous use; maintenance is challenging for big, fast data, which is why we want automatic support in modern industrial applications. A better than the traditional way is to construct order relations automatically when the data attributes to be fused are inspected.

Data fusion builds data sets (or families of data sets). The next step is to extract process *information* from the (often big) data sets with analytics methods such as data mining, statistical analysis, machine learning, computational intelligence, visualization, etc. The process is continued with analytical techniques to fuse sets of information to more meaningful summary information—*information fusion* (cf. Carlsson et al. [13–15]). Some early research results show that information fusion reduces the uncertainty in (social) big data by extracting key (valid, relevant) factors, cleaning out outliers, high-lighting illogical assumptions, etc. (cf. Morente-Molinera et al. [30, 31]).

Knowledge fusion builds on first data fusion, then information fusion. Knowledge fusion applies taxonomies or ontologies—in the D2I industry/university research program (Tekes 340/12) fuzzy ontology was developed and used to detect, identify and deal with recurring problems in pump valve packages (Carlsson et al. [13–15]). Automated knowledge builds on natural (or near-natural) language processing, information extraction with analytics tools, information integration (or federation), computational intelligence (soft computing, evolutionary computation, swarm intelligence, intelligent agents, etc.) (cf. Morente-Molinera et al. [30, 31]).

With the proposed basis in digital fusion, we can then look for theory and technology constructs that could be used for the digital coaches.

One of the approaches to *virtual coaches* builds on the emerging technology of embodied conversational agents (ECA's). ECA's are animated virtual characters, displayed on a computer or a mobile device screen. ECA's play the roles of teachers, mentors, advisors, social companions, and, increasingly, of virtual coaches (cf. Hudlicka [23]). The ECA's engage in natural interaction with humans through dialogue and non-verbal expression which requires minimal or no training; we will probably not be aided by animated virtual characters but the core agent constructs may be useful for the digital coaching when working on data and information fusion material. With some quick sketching we could work with the following setup and functionality: the {$Coach_i$} family will guide users in menu hierarchies and functions of the GUI.

The *virtual trainer* systems are becoming popular as supporting services to fitness and wellness applications; they are typically identified as three classes (i) smart phone applications, (ii) sensor devices, and (iii) image processing devices. Sensor data and images are collected and processed through data fusion, which can be a low level implementation that builds on fast, efficient process monitoring. In our present context, this could be used for working out the parameters needed for the Markov processes (cf. Sect. 3.1). The "trainer" gives feedback on the progress of the exercise and offers summary post exercise data for learning and for motivation to keep up the exercises; thus the virtual trainer type of coaching would be useful also for the GUI support.

3.1 Coaching with Markov Decision Processes

Fern et al. [18] work out a theory base for personalised AI systems that work as personal assistants to support human users with tasks they do not fully know how to carry out. This type of technology has gained much attention in the last 10 years because of the growing use of automated systems with intelligent functions. Fern et al. [18] work out a model where the assistant (an AI system) observes the user (represented as a goal-oriented agent) and must select assistive actions from a closed set of actions in order to help the user achieve his goals.

The interesting thing is that this functionality builds on a decision-theoretic model, which is worked out with partially observable Markov decision processes (POMDPs). Fern et al. [18] work out variations of these Markov processes to get a formal basis for designing intelligent assistants. A specific case is the hidden goal Markov decision processes (HGMDP) that are designed to cover the application environment and the user's policy and hidden goals. The HGMDP is a tuple < S, G, A, A', T, R, π, I_S, I_G > , where S is a set of states, G is a finite set of possible user goals, A is a set of user actions, A' is a the corresponding set of assistant actions, T is a transition function that decides the transition from s to s' (element of S) after the user takes action a (element of A) towards a goal g (element of G); R is a reward function for both the user and the assistant, π is the user's (optimal) policy mapping to the context, and I_S the initial and I_G the goal states. Markov processes are generic constructs that can be used to describe complex processes in a fairly compact form. The digital coaching, we want to build, appears not to be neither an POMDP nor an HGMDP because (A, A') are not stochastic but decided by the application logic and the fuzzy ontology reasoned; the R could be a (fuzzy) distance function.

Some useful constructs build on the Fern et al. [18] theoretical framework. The first is a special case where the assistant's policy is deterministic for each specific goal. This opens up for the use of an optimal trajectory tree (OTT) where the nodes represent the states of the MDP reached by the prefixes of optimal action sequences for different goals starting from the initial state. Each node in the tree represents a state and a set of goals for which it is on the optimal path from the initial state. The size of the optimal trajectory tree, which needs to be reasonably compact for computational purposes, is bounded by the number of goals times the maximum length of any trajectory, which is at most the size of the state space in deterministic domains. This gives some hints at what constructs to look for when trying to work out the {Coach$_j$} and {Coach$_k$} families.

Another interesting result is the approach to solve the problem of selecting an assistive action. For an HGMDP Fern et al. [18] work out a combination of bounded look-ahead search and myopic heuristic computations (selecting an action that has the highest probability of being accepted). By increasing the amount of look-ahead search, the actions returned will be closer to optimal at the cost of more computation; for many HGMDPs the useful assistant actions can be computed with relatively little or no search.

We will need some constructs to learn (e.g.) an HGMDP while interacting with the context, i.e. the assistant (when implemented as {Coach$_j$} and {Coach$_k$}) should follow how the user interacts with the context and learn the user's policy and goal distributions. These constructs would be useful when the assistant is called upon many times for the same construct (quite often at irregular intervals) by the same user; a further extension would be for the assistant to start from basic constructs obtained with one user and then to learn another user's policy and goal distributions. The classical approach is to use Maximum Likelihood estimates of the user's policy distributions from continuous follow-ups and combine that with estimates of the goal attainment (e.g. as fuzzy distances). Another approach that Fern et al. [18] propose is to use an MDP model of the context and bootstrap the learning of the user policy. This would be useful if the user is near optimal in his policy choices and will likely select actions that are near optimal for a selected goal and an actual context.

3.2   Coaching with Virtual Environments

Fricoteaux et al. [19] work out the use of virtual environments for fluvial navigation; these environments offer training in easily modifiable environmental conditions (wind, current, etc.), which have impacts on the behaviour of a ship; fluvial
navigation would in our context be applied to quickly occurring and changing
problem situations. The main difficulty in fluvial navigation is to anticipate
manoeuvres and the variability of the conditions of the environment. It is interesting
to note that the formal framework to represent and support the decision-making
system builds on classical Dempster-Shafer theory in order to take account of
uncertainty. Unlike the theory of probability, the DS-theory allows for explicit
modelling of ignorance. This can be combined with directed graphs to represent
influences between variables; if the inference is probabilistic, Bayesian networks
can be used; with belief functions there are evidential networks with conditional
belief functions (ENCs). Then in turn, ENCs have been generalised by evidential
networks with conditional belief functions (DEVNs), etc.—thus there appears to be
constructs available to apply to the building of digital coaches. The remaining
challenge appears to get it done.

Bloksmal and Struik [4] work out a program for coaching farmers using human
health as a metaphor for farm health, which helps both them and the farmers to gain
an understanding of the issues that are crucial for improving the processes and the
productivity of a farm (which may be far from our context of paper machines). The
coach and the farmer together work out the course of life of the farm, they learn
from what has happened in the history of the farm and translate images of possible
futures into the current state of the farm (not unlike tracing possible consequences
of incidents or problems in parts of the paper machine). The coach operates like a
physician—"alternatingly observing the diseased part and the whole being of an ill
person"—and by referring to this metaphor, opens up similar mechanisms for what
may be wrong with the farm (cf. parts of a paper machine). If done skilfully, it will
show to what extent the farm resembles a living and healthy entity and the farmer
will get new ideas on how to improve "the living farm organism" (cf. the interaction
of parts and the operational characteristics of the paper machine). The use of
metaphors takes out the blame from the narrative; the farmer (cf. the paper machine
operator) will not feel that the coach blames him for having done something wrong.

Bloksmal and Struik [4] show that the process to find and describe the identity
and key processes of a farm is not easy; they use a narrative method—the coach
listens to discover the drama behind the facts. This approach could work for our
digital coaching of knowledge mobilization—i.e. the event reports search and
retrieval tools could be good representations of the "drama behind the facts" and
help to get difficult problems resolved.

The digital coaching is an obvious response to the introduction of advanced and
complex analytical tools, which will be needed to cope with the complexities of
large, automated industrial processes. The state-of-the-art seems to show that there
are fruitful issues for effective research.

4 Summary and Future, Next Scenarios

The theory framework built by and for the classical Operations Research and Management Science since the early 1950s is now in a process of transformation by (Business) Analytics, which is getting the attention of major corporations and senior management. The research groups working on fuzzy multiple criteria optimisation have been part of this process and have specialised in working out and using analytics methods, which implement soft computing theory and algorithms. This has turned out to be very effective and useful for planning, problem solving and decision making in "big data" environments; the "big data" is one of the challenges of the modern digital economy and for which we propose that fuzzy sets would offer instruments for breakthrough research.

Analytics adds value to management; it promotes data-driven and analytical decision-making and "reinvented" fact-driven management. Analytics builds on recent software improvements in information systems that has made data, information and knowledge available in real time in ways that were not possible for managers only a few years ago (Davenport and Harris [17]).

In the context of the digital economy, common wisdom finds that real-time management is a necessity as operations should be planned and carried out in a fast changing and complex environment where careful and thoughtful management will be bypassed by fast, innovative approaches (which may turn out to be of inferior quality, but have then already established sustainable competitive positions). On the other hand, as we have shown with our paper machine example, when there is a problem in a major production process fast reaction and action are priorities; the costs of the damage rises very quickly and can grow very large.

Real-time management is challenged by "big data" and the necessity for fast processing using advanced analytics methods. The advanced analytics would require postdoc-qualified managers—these are rather scarce in senior management positions. Thus there will be a need to reinstate "coaching" functions with the advanced analytics methods to tell/explain to the users what can/should be done, how it should be carried out, what the results are and what they mean, and how they should be applied (with explanations of what could/should not be done).

We have worked out a context and a scenario for digital coaching with the help of a case description of how problems with a paper machine could be handled with a knowledge mobilisation process. The knowledge mobilisation works on event reports about how previous problems were solved and what the best approaches were to finding solutions. The event reports are numerous (tens or hundreds of thousand) and are classified with systems of keywords. The keywords are sometimes overlapping, incomplete or imprecise ("real experts will identify them") which made classical retrieval methods impractical and not very productive. We developed a fuzzy ontology for keyword identification and retrieval and found out that this approach is more productive than the classical methods. Then it turned out that using the fuzzy ontology tools was challenging for the operators in the monitoring and control room for the paper machine.

We discussed the need for coaching in the industrial setting, but quickly found out that it would be both expensive and impractical to try to find and use experienced human coaches. The alternative is online digital coaching and we only need to find some effective methods to build and implement digital coaches. We found out that there are not many useful approaches offered in the literature, the closest we could get was an elaborate framework built around Markov decision processes, which was not actually up to the task. We will work out some possible theoretical frameworks for the digital coaches in a coming series of papers.

We will conclude with a case story of why it will make sense to work with analytical models even in a context, which is not recognized as a domain for analytics. Kahneman [24] relates the case of Orley Ashenfelter, a Princeton economist and wine lover, who wanted to find a way to predict the future value of fine Bordeaux wines from information available in the year they are made (cf. [24]). He was of course well aware that he was stepping on the sensitive toes of world-renowned experts who claimed that they could predict the value development for individual wines over years to come and also in which year the wines will reach their peak quality and highest price. The experts built their judgement on tasting the wines and decades of experience of and insight in the wine markets; Ashenfelter built his predictions on regression analysis and an effective use of statistics tools—he had no possibility to actually taste the wines. Ashenfelter collected statistics on London auction prices for select mature red Bordeaux wines 1990–1991 (sold in lots of a dozen bottles); mature red Bordeaux were defined as vintages 1960–1969 and the wines selected came from six Bordeaux chateaux which are large producers with a reputation to have produced high quality wine for decades—or centuries in some cases. He thus made certain to have a large enough, consistent data set. Ashenfelter found out that the quality of the Bourdeaux wines is decided by (i) the age of the vintage, (ii) the average temperature over the growing season (April–September), (iii) the amount of rain in September and August (less rain gives better wine), and (iv) the amount of rain preceding the vintage (October–March). These factors are well known and often repeated; they have the benefit that they are all measurable and built on published and easily verifiable facts; Aschenfelter collected data on the vintages 1952–1980 and built a regression model with the four factors which turned out to explain about 80% of the variation in the average price of Bordeaux wine vintages. *His point is, that the future quality of Bordeaux wines can be worked out without tasting the wines or introducing any kind of subjective judgements.* He used his models to predict the price development for new vintages of Bordeaux (the correlation between prediction and actual prices is above 0.90) which he has shared with a crowd of followers that are investing in promising, good vintages. Through his models, he has found a few very good vintages that are under-priced in the market. He and his friends have invested in these wines and much enjoyed them.

First, we can notice that Aschenfelter made sure that he had observations on large selections of wine over 10 years from six large chateaux—but only in Bordeaux in order to reduce the number of external factors that influence the wine production but are not relevant for the key issues of his study. Second, his models forecast future prices (years, and even decades into the future) more accurately than the current market prices of young wines do that build on expert estimates; this challenges economics theory that claims that market prices (in effective markets) will reflect all information on the products. Third, experts make judgements that are inferior to algorithms; Kahneman (cf. [24]) argues that some reasons for this is that experts try to be clever, to think outside the box and to work with (too) complex combinations of features to make their predictions. Complexity may work in specific cases but will reduce validity in most cases. There is a *second point* to be made—*analytics, when the proper methods are developed and used, will give insight that intuition and experience will not be able to produce*. This is a lesson learned for the handling of problems (and sometimes disasters) in large industrial production processes when it is claimed that the need to make (almost) real time decisions makes it necessary to forego analytics and rely on the intuition and experience of seasoned operators and engineers rushing to the rescue.

References

1. Acampora G, Loia V (2005) Fuzzy control interoperability and scalability for adaptive domotic framework. IEEE Trans Industr Inf 1(2):97–111. doi:10.1109/-TII.2005.844431
2. Acampora G, Gaeta M, Loia V, Vasilakos A.V (2010) Interoperable and adaptive fuzzy services for ambient intelligence applications. ACM Trans Auton Adapt Syst 5(2):8:1–8:26. doi:10.1145/1740600.1740604
3. Bellman RE, Zadeh LA (1970) Decision-making in a fuzzy environment. Manage Sci Ser B 17(1970):141–164. doi:10.1287/mnsc.17.4.B141
4. Bloksmal JR, Struik PC (2007) Coaching the process of designing a farm: using the healthy human as a metaphor for farm health. NJAS 54–4(2007):413–429
5. Bordogna G, Pasi G (2010) A flexible multi criteria information filtering model. Soft Comput 14:799–809. doi:10.1007/s00500-009-0476-3
6. Borgonovo E, Peccati L (2004) Sensitivity analysis in investment project evaluation. Int J Prod Econ 90:17–25
7. Carlsson C, Brunelli M, Mezei J (2010) Fuzzy ontology and information granulation: an approach to knowledge mobilisation. In: International conference on information processing and management of uncertainty in knowledge-based systems (IPMU 2010), June 28–July 2, 2010, Dortmund, Germany, In: Hüllermeier E, Kruse R, Hoffmann F (eds) Information processing and management of uncertainty in knowledge-based systems, vol 21. Springer, Berlin, Heidelberg, [ISBN: 978-3-642-14057-0], pp 420–429. doi:10.1007/978-3-642-14058-7 44
8. Carlsson C (1984) On the relevance of fuzzy sets in management science methodology. In: Zimmermann HJ, Zadeh LA, Gaines BR (eds) Fuzzy sets and decision analysis. TIMS studies in management sciences, vol 20, North-Holland, Amsterdam, pp 11–28
9. Carlsson C, Fullér R (2001) On possibilistic mean value and variance of fuzzy numbers. Fuzzy Sets Syst 122:315–326
10. Carlsson C, Fullér R (2002) Fuzzy reasoning in decision making and optimization. Springer, Berlin, Heidelberg
11. Carlsson C (2012) Soft computing in analytics: handling imprecision and uncertainty in strategic decisions. Fuzzy Econ Rev XVII(2):3–12

12. Carlsson C, Heikkilä M, Mezei J (2014) Possibilistic bayes modelling for predictive analytics. In: Proceedings of 15th IEEE international symposium on computational intelligence and informatics. Budapest, Nov 2014, pp 15–20
13. Carlsson C, Mezei J, Wikström R. (2015) Aggregating linguistic expert knowledge in type-2 fuzzy ontologies. Appl Soft Comput 35:911–920
14. Carlsson C, Heikkilä M, Mezei J (2016) Fuzzy entropy used for predictive analytics. In: kahraman C (ed) Fuzzy sets in its 50th year. new developments, directions and challenges, studies in fuzziness, vol 341, Springer, pp 2–3
15. Carlsson C (2016) Imprecision and uncertainty in management—the possibilities of fuzzy sets and soft computing. NOEMA XV, Rom Acad Sci 2016:89–114
16. Cross V (2004) Fuzzy semantic distance measures between ontological concepts. In: IEEE annual meeting of the north American fuzzy information processing society (NAFIPS 2004), Banff, AB, Canada, June 27–30
17. Davenport TH, Harris JG (2007) Competing on analytics. Harvard Business School Press, Boston, The New Science of Winning
18. Fern A, Natarajan S, Judah K, Tadepalli P (2014) A Decision-theoretic model of assistance. J Artif Intell Res 49:71–104
19. Fricoteaux L, Thouvenin I, Mestre D (2014) GULLIVER: a decision-making system based on user observation for an adaptive training in informed virtual environments. Eng Appl Artif Intell 33(2014):47–57
20. Hirvonen J, Tommila T, Pakonen A, Carlsson C, Fedrizzi M, Fullér R (2010) Fuzzy keyword ontology for annotating and searching event reports. In: International conference on knowledge engineering and ontology development (KEOD 2010), Valencia, Spain, October 25–28, 2010
21. Holi M, Hyvönen E (2005) Modeling degrees of overlap in semantic web ontologies. In: Proceedings of the ISWC workshop uncertainty reasoning for the semantic web, Galway, Ireland. http://www.seco.hut.fi/publications/2005/holi-hyvonen-modelingdegrees-of-overlap-2005.pdf
22. Holi M (2010) Crisp, fuzzy and probabilistic faceted semantic search, doctoral dissertation, Aalto University School of Science and Technology, [ISBN 978-952-60-3183-5], http://lib.tkk.fi/Diss-/2010/isbn9789526031842/
23. Hudlicka E (2013) Virtual training and coaching of health behavior: example from mindfulness meditation training. Patient Educ Couns 92(2013):160–166
24. Kahneman D (2011) Thinking fast and slow. Penguin Books, London
25. Lee C-S, Jian Z-W, Huang L-K (2005) A fuzzy ontology and its application to news summarization. IEEE Trans Syst Man Cybern B Cybern 35(5):859–880. doi:10.1109/TSMCB.2005.845032
26. Lee CS, Wang, G, Acampora M-H, Hsu C-Y, Hagras H (2010) Diet assessment based on type-2 fuzzy ontology and fuzzy markup language. Int J Intell Syst 25(12):1187–1216. doi:10.1002/int.20449
27. Liberatore M, Luo W (2011) INFORMS and the analytics movement: the view of the membership. Interfaces 41(6):578–589
28. Majlender P (2004) A normative approach to possibility theory and soft decision support, TUCS Dissertations, 54, Turku
29. Mezei (2011) A quantitative view on fuzzy numbers. TUCS Dissertations, 142, Turku
30. Morente-Molinera, JA, Wikström R, Carlsson C, Viedma-Herrera E (2016) A linguistic mobile decision support system based on fuzzy ontology to facilitate knowledge mobilization. Decis Support Syst 1
31. Morente-Molinera JA., Mezei J, Carlsson C, Viedma-Herrera E (2016) Improving supervised learning classification methods using multi-granular linguistic modelling and fuzzy entropy. Trans Fuzzy Syst
32. Straccia U (2010) SoftFacts: a top-k retrieval engine for ontology mediated access to relational databases. In: IEEE international conference on systems man and cybernetics (SMC), pp 4115–4122. doi:10.1109/ICSMC.2010.5641780

33. Tommila T, Hirvonen J, Pakonen A (2010) Fuzzy ontologies for retrieval of industrial knowledge—a case study. In: VTT working papers, number 153/2010, http://www.vtt.fi/inf/pdf/workingpapers/2010/W153.pdf [ISBN 978-951-38-7494-0]
34. Zadeh L (1965) Fuzzy sets. Inf Control 8(3):338–353

Printed in the United States
By Bookmasters